An introduction to science studies
the philosophical and social aspects of science and technology

JOHN ZIMAN FRS

Published by the Press Syndicate of the University of Cambridge
The Pitt Building, Trumpington Street, Cambridge CB2 1RP
40 West 20th Street, New York, NY 10011-4211, USA
10 Stamford Road, Oakleigh, Melbourne 3166, Australia

© Cambridge University Press 1984

First published 1984
First paperback edition 1987
Reprinted 1988, 1992, 1994, 1995

Printed in Great Britain by
Athenæum Press Ltd, Gateshead, Tyne & Wear

Library of Congress catalogue card number 84-7830

British Library cataloguing in publication data
Ziman, John
An introduction to science studies
1. Science – Social aspects
1. Title
306'.45 Q175.5
ISBN 0 521 34680 0 paperback

Contents

	Preface		ix
1	'Academic' science		1
	1.1	Different aspects of science	1
	1.2	The chain of discovery	2
	1.3	'Internal' and 'external' sociologies of science	3
	1.4	Three dimensions of 'academic' science	6
	1.5	Academic science as 'public knowledge'	9
2	Research		13
	2.1	Scientific knowledge	13
	2.2	Description	14
	2.3	Generality	15
	2.4	Patterns of fact	16
	2.5	Investigation	18
	2.6	Instrumentation	19
	2.7	Measurement	20
	2.8	Experiment	22
	2.9	Scientific laws	23
	2.10	Explanation	24
	2.11	Cause and effect	25
	2.12	Models	26
	2.13	Theory	28
	2.14	Hypotheses	29
	2.15	Problem-solving and the growth of knowledge	31
3	Validity		34
	3.1	Epistemology	34
	3.2	Empiricism	35
	3.3	Phenomena and sense-data	37

	3.4	The problem of induction	40
	3.5	Inference	41
	3.6	Prediction	43
	3.7	The hypothetico-deductive method	46
	3.8	Established knowledge	48
	3.9	Does science describe reality?	52
	3.10	Regulative principles of scientific work	55
4		Communication	58
	4.1	The archival literature of science	58
	4.2	Linkage by citation	60
	4.3	What does a scientific paper say?	61
	4.4	How do scientific papers get published?	62
	4.5	Selection by peer review	64
	4.6	The accreditation process	65
	4.7	'Informal' communication between scientists	67
5		Authority	70
	5.1	Recognition	70
	5.2	Exchange of gifts – or competition?	72
	5.3	Specialization	74
	5.4	Invisible colleges	75
	5.5	Stratification	76
	5.6	Functions and dysfunctions of authority	78
6		Rules and norms	81
	6.1	Behaving as a scientist	81
	6.2	The Mertonian norms	84
	6.3	An ethos of academic science	86
	6.4	Does academic science have an ideology?	87
7		Change	91
	7.1	Cognitive change	91
	7.2	Institutional change	93
	7.3	Change by revolution	94
	7.4	The historical structure of scientific revolutions	96
	7.5	The sociodynamics of scientific life	99
8		The sociology of scientific knowledge	102
	8.1	Science and the sociology of knowledge	102
	8.2	Epistemological relativism	103
	8.3	The 'strong programme' in the sociology of knowledge	105

	8.4	Science as a social enterprise	106
	8.5	Establishing a consensus	108
	8.6	Sociological epistemology	109
9		Science and technology	112
	9.1	Science as an instrument	112
	9.2	Science-based technology	113
	9.3	Technology-based sciences	114
	9.4	Scientific technique	114
	9.5	Science *or* technology	115
	9.6	Science *from* technology?	116
	9.7	'S & T'	118
10		Pure and applied science	121
	10.1	'R & D' in 'S & T'	121
	10.2	Growth	123
	10.3	Amateurism and state patronage	123
	10.4	The rise of academic science	124
	10.5	The external relations of academic science	126
	10.6	Industrial science	127
	10.7	Pure science – and its applications	129
11		Collectivized science	132
	11.1	Societal demand	132
	11.2	Apparatus	134
	11.3	Sophistication and aggregation	136
	11.4	Collaboration	137
	11.5	The collectivization of science	138
12		R & D organizations	140
	12.1	Science as an instrument of policy	140
	12.2	The spectrum of relevance	141
	12.3	The philosophy and methods of R & D	143
	12.4	The management of R & D	144
	12.5	The internal sociology of collectivized science	145
13		The economics of research	149
	13.1	Costing the benefits	149
	13.2	Macroeconomics of R & D	150
	13.3	The sources of invention	152
	13.4	The microeconomics of research	154
	13.5	Economic incentives for R & D	156

14	Science and the State	159
	14.1 Government support for science	159
	14.2 The politics of science	160
	14.3 Criteria for choice	161
	14.4 The dilemma of patronage	164
	14.5 The limits of control	166
	14.6 Science in government	169
15	The scientist in society	173
	15.1 Towards a social psychology of science	173
	15.2 The scientist as intellectual entrepreneur	174
	15.3 Citizen of the republic of science	176
	15.4 The scientist as technical worker	177
	15.5 The scientist as expert	178
	15.6 Social responsibility in science	180
16	Science as a cultural resource	183
	16.1 Beyond the instrumental mode	183
	16.2 Public understanding of science	184
	16.3 Folk science, pseudo-science and parascience	185
	16.4 Academic scientism	187
	16.5 Science and values	190
	16.6 The value of science	192
	Index	195

Preface

In *Teaching and Learning about Science and Society* (Cambridge University Press, 1980), I argued at length that everybody ought to learn something *about* science, but that science is a large and open-ended topic, which needs to be treated in various ways at various stages of educational maturity. At school level, the most natural approach is through case studies of the place of science and technology in modern life, as we presented them, for example, in the *SISCON in Schools* units (published in 1983 by the Association for Science Education and Basil Blackwell). For slightly older students, a conception of science as a social institution can be built up from historical case studies, along the lines of the lectures I wrote up as *The Force of Knowledge* (Cambridge University Press, 1976).

The present work goes one level deeper. It is addressed to students – and other diligent readers – who want to discover, beneath the historical and contemporary particulars, a more general framework of principle. They want to understand what is being said about science by the historians, philosophers, sociologists, psychologists, economists and political scientists who have been making such notable contributions to 'science studies' these last few years. They need access to the scholarly literature in these various fields, both for its intrinsic interest and as a possible guide to action in scientific research, in industrial management, in political administration, and in public affairs.

Each of these fields has its own basic textbooks, 'readers' and advanced treatises. But there is a natural tendency, in each case, to look upon the subject from the viewpoint of a particular discipline, and to overelaborate the features that are mainly visible in that aspect. The student is seldom shown how these features might appear from other points of view, and thus never gets a coherent impression of the subject as a whole. In many cases, also, the most instructive writings start at quite a high scholarly level, making it difficult for the beginner to appreciate what is really at stake.

This book actually arose out of a course of lectures which students of physics, philosophy or sociology at Bristol University could take as one of the examined options in their bachelors' degrees. What seemed to be needed at that stage was a unified account, in the plainest possible language, of the general concepts and significant issues in this interdisciplinary area. In effect, it is an elementary treatise on *metascience* – the 'science of science' in the broadest sense – intended for use as a basic text in specialized undergraduate and postgraduate courses in all fields of science studies. It cannot pretend to be definitive in any one field, but shows by its approach and by many cross references from chapter to chapter, the relationships that exist between these fields.

A work of this kind craves infinite charity from scholars with specialized knowledge of particular topics. This charity is begged, not only for errors of fact or principle but also for apparently neglecting many valuable insights from many distinguished contributors. I know how much I myself owe to the evocative writings and distinctive ideas of a number of brilliant scholars in this field and, as the reading lists indicate, I should like every student to enjoy them and benefit from them as I have. But it seemed more useful to present the essential themes in my own language, and on my own terms, as a sympathetic rapporteur, rather than putting together a pastiche of other people's opinions in a medley of discordant voices. This means, for example, that I have tended to approach each topic in the first instance from the naturalistic standpoint of a working scientist or science student, and then to move round to a more philosophical or sociological stance as the analysis develops. I have also taken a personal line in drawing attention to the 'model' of academic science that I set out in detail in *Public Knowledge* (Cambridge University Press, 1967) and *Reliable Knowledge* (Cambridge University Press, 1978) and have devoted a good deal of space to the 'collectivization' of science in recent years, which seems to me to be a much more significant phenomenon than most other observers would allow. But this book is idiosyncratic only as an attempt to make sense of a very complicated and loosely articulated body of knowledge, and the reader should not take at face value the confidence with which some of my opinions are apparently expressed. Remember, please, that this is only an *introduction* to science studies, not an authoritative account of what is known. To make proper use of it, the student should read as deeply as possible into the recommended works, which will draw him or her further into 'fresh woods and pastures new'.

The writing of the final text during the past 18 months has been greatly facilitated by the generous award of a Visiting Professorship at this College. I am especially indebted to Elspeth Robinson and Joan Wright, who typed the manuscript for me. But work on this book really began many years ago, when I first prepared notes for these lectures, and it continued as I revised them in later years. All that time I had the good fortune to be a member of the H. H. Wills Physics Laboratory at Bristol, a particularly happy and distinguished university department. This is the moment to express my gratitude to all my colleagues there, not only for their unalloyed personal friendship but also for the support and encouragement they gave me in this novel educational enterprise. And now, having had to read up and write up the philosophy chapters for myself, I realize how much our students owed to Stephan Körner and David Hirschman, who taught this part of the course with such skill and understanding.

John Ziman
Department of Social & Economic Studies,
Imperial College of Science & Technology,
London
December, 1983

1
'Academic' science

'Such...is the respect paid to science that the most absurd opinions may become current, provided they are expressed in language, the sound of which recalls some well-known scientific phrase.'
James Clerk Maxwell

1.1 Different aspects of science

What *is* 'Science'? Our whole approach to the subject of this book depends on how we might be tempted to answer this question. But it is really much too grand a question to be answered in a few words. Conventional definitions of science tend to emphasize quite different features, depending upon the point of view. Each of the metascientific disciplines – the history of science, the philosophy of science, the sociology of science, the psychology of creativity, the economics of research, and so on – seems to concentrate upon a different aspect of the subject, often with quite different policy implications.

For example, if science is defined as 'a means of solving *problems*', this emphasizes its *instrumental* aspect. Science is thus viewed as closely connected with *technology*, and hence an appropriate subject for *economic* and *political* study. The implication that this instrument should be used wisely and well puts it into the open arena of social conflict.

Another definition of science – as 'organized *knowledge*' – emphasizes its *archival* aspect. Information about natural phenomena is acquired by research, organized into coherent theoretical schemes, and published in books and journals. Although this knowledge is often profoundly influential through its technological applications, there is much to be said for treating it as a politically neutral, public resource. The accumulation of scientific knowledge is thus a significant *historical* process, worthy of special study.

Or we may follow an old *philosophical* tradition by emphasizing the *methodological* aspect of science. Procedures such as experimentation, observation and theorizing are considered elements of a special *method* for obtaining reliable information about the natural world. From this point of view, science may be regarded as essentially objective, and hence transcending all political considerations.

Finally, one might emphasize the *vocational* aspect of science by tacitly defining it as 'whatever is discovered by people with a special gift for *research*'. This draws attention to such important personal aptitudes as curiosity and intelligence, which are well worth *psychological* investigation. Such studies might suggest that scientists should be recognized as members of a distinct *profession*, of considerable political significance.

There is so much that can be said about science from each of these and from other aspects that there is a tendency within each metascientific discipline to treat its own special definition as self-sufficient. Thus, philosophers of science largely ignore its instrumental and vocational features, whilst many serious studies of the political role of science seem quite oblivious to its complex methodological and vocational aspects. It is instructive to read books about science in this light. It almost seems as if each discipline has in mind a different 'model' of science, constructed around just those particular features in which it happens to be interested.

In truth, science is all these things, and more. It is indeed the product of research; it does employ characteristic methods; it is a body of organized knowledge; it is a means of solving problems. It is also a social institution; it needs material facilities; it is an educational theme; it is a cultural resource; it requires to be managed; it is a major factor in human affairs. Our 'model' of science must relate and reconcile these diverse and sometimes contradictory aspects.

1.2 The chain of discovery

The four conventional definitions of science are obviously complementary, but how should they be connected? They are often supposed to fit together in a one-way chain, from the vocational aspect, through the methodological and archival aspects, to the instrumental aspect where science merges into technology. That is to say, scientific knowledge is generated by individual scientists in the form of *discoveries*, which must be validated by scientific methods before being published in archival form. This knowledge is then applied to the solution of whatever problems may have arisen in society (fig. 1).

The great advantage of the linear *discovery model* of science is that it divides the labour equitably between the major metascientific disciplines. At each stage, so to speak, the material is processed according to the principles of the corresponding discipline and passed on to the next stage. Factors that operate at one stage can be ignored further along the line: 'intuition' for example can be treated as an important factor in the context of discovery (chapter 2) but not in the context of validation (chapter 3) where 'logic' is supposed to rule supreme. In principle, the history of

1.2 'Internal' and 'external' sociologies of science

Fig. 1 The chain of discovery

science ought to deal with *all* stages in this process; in its narrowest traditional practice, it was often little better than a chronicle of publications, without significant reference to their particular psychological sources or general social context.

But the discovery model is really much too simple, for it obviously neglects some very significant realities. It assumes, for example, that information flows only one way along the chain, as if there were no technological demands on basic scientific research. It also takes no account of the fact that scientists do not work alone; research is to a large extent a communal endeavour, where individual action is strongly influenced by social goals and norms.

These are not just minor deficiencies that can be made good later by more detailed analysis. In the last 20 years the whole field of science studies has been transformed by the realization that science can only be understood if it is treated as a *social institution*, both within its own sphere of activity and in its relationships with the world at large. In other words, the *sociology* of science must be included in this programme of study, along with more traditional metascientific disciplines such as philosophy and history.

1.3 'Internal' and 'external' sociologies of science

The programme of study outlined in this book is much more ambitious than a link-by-link journey along the discovery chain, for it implies a much more complicated conception of science as a whole. For clarity of exposition, it is convenient to proceed through the analysis in two successive stages, following the academic convention of distinguishing between the *internal* and *external* sociologies of science. That is to say, in chapters 2 to 8 we study the relationships between scientists as they go about research, and then in later chapters show how scientific work relates to the broader social context in which it is undertaken.

This order of exposition has been chosen solely for simplicity; it does not imply that 'internal' factors are somehow more important than 'external' factors. The reader who perseveres to the end of this book may well come out with quite the opposite view, which emphasizes the primacy of the whole social order, of which science is only one component.

Our starting point, therefore, is a 'model' of science where external forces are entirely neglected. In the traditional scheme, this would be equivalent to cutting the

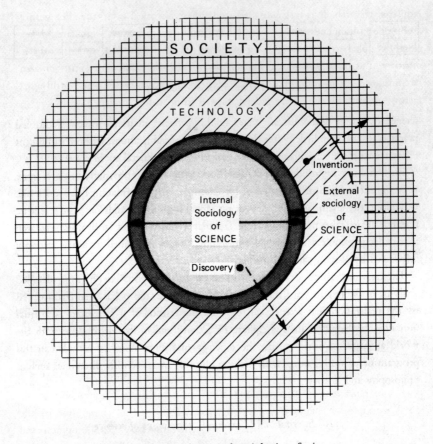

Fig. 2 Internal and external sociologies of science

chain of discovery between its archival and instrumental elements, as if scientific knowledge were accumulated solely 'for its own sake', without any thought for its possible applications. The boundary between 'science' and 'society' is envisaged as a semi-permeable membrane, through which knowledge only flows outward, from the scientific into the technological sphere (fig. 2). The goal of the 'internalist' programme is to account for what goes on in the region bounded by this membrane, philosophically, sociologically and psychologically, without reference to a wider world.

Throughout this book, we shall refer to this as the *academic* model of science. Although it is very far from realistic as a model of contemporary science, it is the notion of the nature of their activities that many scientists, and some metascientists, still have. It also has an important *historical* significance, for it might be considered a fair description of the natural sciences of, say, a century ago, before the rise of industrial research. It is instructive to study these historical cases, together with a few modern disciplines, such as cosmology or pure mathematics, which are not yet closely coupled to technologies, for evidence on the way such a system works in practice: it is often possible to explain the actual behaviour of scientists in such circumstances by an appropriate combination of cognitive, personal and communal factors, almost as if they were indeed isolated from society at large.

Academic Science is thus the characteristic model for the 'internal' sociology of science. In the 'external' sociology of science, on the other hand, the usual assumption seems to be that science is a 'black box', whose internal mechanisms can be ignored. Study is concentrated on the technological effects of knowledge that has percolated outward from 'pure' science, through the 'membrane', and then been applied to the solution of practical problems. The *instrumental* capabilities of science in the service of political, military or commercial forces are thus regarded as paramount. This 'externalist' conception of *industrial science* as primarily a component of *technology* is developed in chapters 9 and 10.

In the final analysis, however, these elementary 'internalist' and 'externalist' accounts of science and technology must be reconciled and properly connected. This calls for a complete revision of both the academic and industrial models. As the sociologists of knowledge have demonstrated from historical research (chapter 8), the 'membrane' separating science from society is largely an illusion; the influences that are always flowing across this mythical boundary have profound effects on either side. It is a cliché that these influences have become so powerful in recent years that science is transforming society around us. What is not always realized is that the inner workings of science itself are being changed out of all recognition by the enormous social forces acting on it, and penetrating to its philosophical and psychological core. The theme of chapters 11 and 12 is the *collectivization* of science into a system of *Research and Development* organizations, whose economic and political characteristics are taken up in chapters 13 and 14.

The subject matter of modern metascience cannot therefore be considered static. The disciplines brought together under the heading of 'science studies' are concerned with a dynamical system that is undergoing dramatic historical change whilst we study it. This change is taking place both within science and in its cultural context. Thus, for example, as indicated in chapters 15 and 16, scientists are now expected to play a much wider variety of social rôles, and society interacts with science in many more ways than in any previous culture. In these final chapters we thus begin to see more clearly some of the contemporary answers to the very first questions posed in this book – what *is* science, and how does it work as a social institution, as a vocation, as a source of belief, and as an instrument of power?

A very important question that might be asked at any stage of the argument is whether the word 'science' is being used in its narrowest or its broadest sense: does it mean the study of 'natural' phenomena by 'objective' techniques, or should it be extended to the interpretation of social systems and psychological events where 'subjective' factors cannot be avoided? For example, should we apply our 'internalist' sociology of science to sociology itself, or discuss the efficacy of 'pure' social psychology in the 'technology' of education? Questions of this kind are certainly of the greatest interest and importance, for they uncover many of the unconscious assumptions that we make when we refer to a body of knowledge as a 'science'. But if we were to try to raise such questions in the earlier chapters of this book, we would probably get some very misleading answers. It seems better to leave this whole issue to the final chapter (§16.4) when we have arrived at a fuller picture of science and its social function, and can decide the most fruitful approach to this very subtle theme. For clarity and simplicity of exposition, almost all the examples of scientific thought and action in the main text are drawn from the 'laboratory' sciences, such as physics, chemistry or biology, or from their technological applications in engineering, medicine, etc., but this is not to be interpreted as an opinion that these are the only genuine scientific disciplines. On the contrary, as argued in §16.4, my own inclination would be towards the wider view, which would include the social and behavioural sciences in the notion of 'science' at every stage in the discussion.

1.4 Three dimensions of 'academic' science

For the moment, however, we are concerned with 'academic' science. Even the most aloof and idiosyncratic pure scientists are not really 'purely seekers after truth'. Their contributions to knowledge are seldom made quite personally and independently of one another. They often work closely together in their research, and almost always refer to themselves as members of an academic *discipline*, and of a corresponding scientific *community*: that is to say, they are quite aware of their social interactions with one another as *scientists*.

These interactions take many different forms. We observe communal *institutions*, such as university departments, learned societies and scientific journals, involved in a variety of communal *activities*, such as scientific education, the publication of scientific papers, debates on controversial scientific questions, or the ceremonial award of prizes for famous discoveries. More abstractly, we notice significant communal *influences*, such as educational curricula, research traditions and research programmes. Every scientist is called upon to play various communal *rôles*, such as graduate student, research supervisor or eminent scientific authority, and is subject to communal *norms* of behaviour, such as 'universalism' or 'disinterestedness'.

Some of these forms of communal interaction have long been obvious. Historians of science have always been interested in the creation of communal institutions and the activities that they foster. No serious history of science in seventeenth-century

1.4 Three dimensions of 'academic' science

Europe, for example, would overlook the foundation of the national scientific academies that brought together so many of the important scientists of their day.

But this interest has tended to be marginal. The fundamental principle of the recent revolution in the metasciences is that these communal institutions, activities, influences, rôles, norms, etc., are not just a background for the logic of scientific method, or to the mystery of scientific creativity; they are *constitutive* of science as we know it. It is not possible to understand the status of scientific theories, or how these get thought of in the first place, without asking how scientists relate to one another in the course of their scientific work.

Any study of the collective actions and relationships between the members of a human group is bound to raise questions about the attitude of the observer and the framework within which the observations are to be interpreted. There are sociologists who insist that this attitude should be as detached and uninvolved as possible, as if one were an anthropologist observing the daily life and occasional festivals of a newly discovered tribe. This *ethnographic* approach has much to recommend it in principle. For the lay outsider, scientific research is an unfamiliar activity imbued with meanings that are only intelligible to the participants, and thus analogous to the symbolic rituals of a mystical sect. Sociological research in this spirit has clearly demonstrated that scientific people and scientific organizations are not at all different, in many essential characteristics, from other people and other organizations of comparable size in comparable cultures.

In practice, however, this very refined approach to human affairs calls for superhuman powers of intellectual detachment and sophistication. The sociology of science is difficult enough to write about in ordinary language, without the additional handicap of trying to purge it of any of the terms that the actors would normally use to describe their own actions to one another. Indeed, there are other schools of sociology that insist that social activity must be interpreted *hermeneutically*, by empathic comprehension of what it means for those taking part. Whatever we may think about this subtle issue in the theory of the social sciences, we are bound to adopt the latter point of view, at least provisionally, in an introductory account of the subject.

The traditional ways of theorizing about academic science usually implied that it had the distinct psychological and philosophical aspects of the discovery model of §1.2. Together with the sociological aspect, these give us three distinct categorial frameworks of abstract description – three different terminologies and conceptual schemes into which the observed phenomena might perhaps be made to fit. The psychology of research uses *personal* terms, such as 'motives', 'perceptions' and 'intelligence'; the philosophy of science uses categories of *knowledge*, such as 'theory', 'contradiction', 'causality'; the sociology of science is about *communities*, with 'institutions', 'norms', 'interests', etc. Each of these schemes has been developed independently in its own 'dimension' up to a high level of intellectual sophistication. The difficulty is that science is a complex activity which exists, so to speak, in all three of these dimensions at once and cannot be understood properly if it is described in three separate 'aspects', without consideration of their interrelations.

Our natural way of talking about science draws indiscriminately on all three schemes. This is very obvious from a simple example, such as the following account of a recent episode, as it might have been reported in a journal such as *Nature* or *Science*:

> 'Darwin's *theory* of evolution by natural selection is widely held to be well established.' (*knowledge*)
>
> 'Nevertheless, at several scientific *meetings* it has come under considerable criticism.' (*community*)
>
> 'According to some recent *experiments* by an Australian scientist, Dr Edward Steele, there is evidence for the inheritance of acquired characteristics.' (*knowledge/person*)
>
> 'Dr Steele was invited to continue this research in the laboratory of Sir Peter Medawar, the world's leading *authority* on immunology.' (*person/community*)
>
> 'There were, however, considerable difficulties over the *publication* of his later results.' (*knowledge/person/community*)
>
> 'Dr Steele was accused of breaking the conventions of scientific *controversy* by making personal attacks on the work of other scientists.' (*knowledge/community*)
>
> 'Although his own *sincerity* in this affair is not in question, the *originality* and professional competence of his experimental research is now seriously doubted.' (*person*)

These sentences evidently make a connected paragraph containing words that belong to all the different schemes or combinations of schemes in this three-dimensional framework (fig. 3). In the most complicated case, we use words that belong to all three schemes at once. Thus, for example, a scientific publication obviously conveys scientific information, and therefore has a cognitive or *philosophical* dimension. At the same time it is addressed to a segment of the scientific community, and therefore has a communal or *sociological* dimension. In addition, there is the *historical* axis, along which science may be said to evolve by a linked sequence of research publications. These 'dimensions', moreover, are not independent of one another: thus, for example, the psychological significance of a scientific paper to its author is closely connected with the philosophical status of the research results it claims. It would be quite misleading to gloss over such connections for the sake of theoretical simplicity.

All human activities have their personal and communal aspects. The complementarity of individual and collective descriptions is a commonplace of social theory.

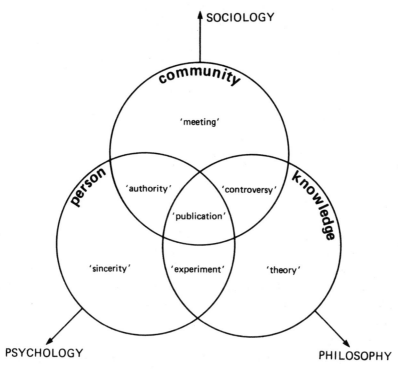

Fig. 3 Three dimensions of discourse about science

The peculiarity of science is the highly ordered and compelling symbolism of the knowledge by which it is both bound together and transformed.

1.5 Academic science as 'public knowledge'

The true realm of discourse about science must surely be of many dimensions. Nevertheless, most of what is really worth knowing in the field of science studies has been established under the banner of one or other of the conventional metascientific disciplines, with little support from the others. Each of the next few chapters could thus be labelled according to the discipline from which it mainly derives. Chapters 2 and 3, for example, are essentially a survey of the epistemological problems normally studied by philosophers of science, whilst chapters 5 and 6 on authority, rules and norms are obviously sociological in language and spirit. A conventional study programme would take up each of these topics in detail, interpreting it as far as possible within the framework of its characteristic discipline. But such a programme leaves one with numerous oversimplifications and inter-

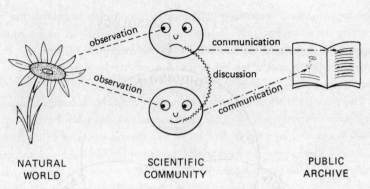

Fig. 4 Academic science as a social system

disciplinary loose ends. Epistemology, for example, is more dependent on sociology than we like to think: in practice the credibility of a scientific conjecture depends very much on the scientific status of the person who moots it! To correct these misconceptions and make good these missing links, we are bound to set up some general 'model' or multidisciplinary interpretative scheme of academic science, in which these diverse topics can be seen to be significant and relevant to the subject as a whole.

At a purely descriptive level, science can obviously be represented naïvely as a community of individual scientists observing the natural world, discussing their discoveries with one another and recording the results in the archives (fig. 4). This model correctly defines research as a social activity, but it lacks any dynamical principle beyond the personal curiosity of its individual members. This sort of scheme can be given much more force and coherence if it is organized around the following proposition:

Academic science is a social institution devoted to the construction of a rational consensus of opinion over the widest possible field.

This is not intended as yet another *definition* of science: it is a hypothetical characterization, of which every word is open to question, criticism and empirical testing. But it does span all the metascientific disciplines, and thus provides an active principle by which to link together many observed features of the 'academic' style of science. The obvious archival, methodological, communal and vocational features of science can be related to it without forcing or major inconsistency, and yet it is relatively undogmatic on the vexed question of the validity or 'truth' of scientific knowledge. It also suggests some very general demarcation criteria by which science can be distinguished from cognate social activities or institutions, such as technology, law, religion, art, education and humanistic studies.

The conception of science as simply 'public knowledge' is, of course, avowedly 'internalist'. It suggests no other goal for the enterprise than the search for knowledge 'for its own sake'. 'Externalist' influences are completely excluded. But this is precisely the attitude we have decided to adopt in the first part of this book. It is necessary to understand the traditional academic conception of science on its own ground, and in its own terms, before we can appreciate how far this might be from present-day reality.

This principle is also a temptation towards *functionalism*. It tends to suggest that the detailed characteristics of science as we happen to know it are somehow essential to its 'functioning' as a whole. One might then use this argument to justify existing practices – for example, the 'peer review' procedure by which anonymous 'referees' scrutinize scientific papers submitted for publication – although they may, in fact, have grown up more or less by accident, and could well be managed quite differently. Social institutions are both more ramshackle and more adaptive to historical change than such arguments allow.

In any case, whether or not one accepts 'maximum rational consensus' as the fundamental objective of academic science, this principle is very convenient as a provisional hypothesis around which to structure one's observations and conjectures about the way in which scientists really work. As in the natural sciences themselves, this is the proper spirit in which to undertake a scholarly investigation.

Further reading for chapter 1

To appreciate the theme of this book, the reader should have access to works on the scope of modern science and on its historical development. It is impossible to make a short list of recommended books on this limitless subject, since the choice must depend on each reader's existing knowledge and interests. For a readable general account, one might turn to

J. D. Bernal, *Science in History*. London: Watts, 1954

although his particular interpretations of historical events and contemporary circumstances are often more controversial than he indicates.

For information on particular topics, one might look first in

H. T. Pledge, *Science since 1500*. London: HMSO, 1966

or C. Singer, *A Short History of Scientific Ideas to 1900*. Oxford: Clarendon Press, 1959

For an account of the formation of the modern scientific attitude, one could start with

H. Butterfield, *The Origins of Modern Science 1300–1800*. London: Bell, 1957

or A. R. Hall, *The Scientific Revolution 1500–1800*. London: Longmans, 1954

These are but entry points to an immense literature, dealing with particular periods, particular scientific disciplines, or the life and works of particular scientists. Books and articles dealing specifically with historical aspects of the social relations of science will be noted in later chapters. A compact, but well-ordered and reasonable survey of all this literature, with special reference to the 'internalist–externalist' debate, is given by

> R. MacLeod, 'Changing Perspectives in the Social History of Science', in *Science Technology and Society*, ed. I. Spiegel-Rösing & D. de Solla Price, pp. 149–96. London: Sage, 1977

An elementary schematic account of 'Science Studies' is given by

> J. M. Ziman, *Teaching and Learning about Science and Society*. Cambridge: Cambridge University Press: 1980 (chapters 5–7)

A more scholarly survey, with full biography, is given by

> I. Spiegel-Rösing, 'The Study of Science, Technology and Society (SSTS): Recent Trends and Future Challenges', in *Science Technology and Society*, ed. I. Spiegel-Rösing & D. de Solla Price, pp. 7–42. London: Sage, 1977

The social model of academic science introduced in §1.5 is discussed at length in

> J. M. Ziman, *Public Knowledge*. Cambridge: Cambridge University Press, 1967 (especially pp. 1–29)

Another elementary model, which emphasizes external factors from the beginning, is given by

> L. Sklair, *Organised Knowledge*. London: Hart-Davis, MacGibbon, 1973 (pp. 57–63)

The most readable example of an ethnographic study of scientific work is

> B. Latour & S. Woolgar, *Laboratory Life*. London: Sage, 1979 (pp. 43–104)

2

Research

'...in the discovery of secret things and in the investigation of hidden causes, stronger reasons are obtained from sure experiments and demonstrated arguments than from probable conjectures and the opinions of philosophical speculators of the common sort.'
William Gilbert

2.1 Scientific knowledge

The purpose of science is to obtain *scientific knowledge*. That is to say, scientific work is directed towards acquiring a special type of information, either for immediate practical use or for publication in textbooks, encyclopaedias, learned journals, etc., under various headings such as physics, chemistry or biology. A typical item of scientific information might be, say: 'The benzene molecule contains six carbon atoms arranged in a ring'. This is clearly somewhat different from the sort of knowledge usually to be found in novels, law reports, sermons, or political manifestos – for example, that 'it is love that makes the world go round', or 'the greater the truth, the greater the libel'.

But what are the distinguishing features of *scientific* knowledge as such? This traditional philosophical question is important because it may decisively affect our actions to know that a particular piece of information is 'scientifically' warranted (cf. §16.3). It is also one of the key questions about science as a human activity, for it asks about the fundamental objectives of research. If, for example, it is said that the goal of science is to make 'discoveries', then we must have some idea what sort of thing a scientific discovery is supposed to be. If, further, we insist that scientific knowledge is only to be gained by a special 'method', then we naturally enquire whether this method makes this sort of knowledge peculiarly valid.

These are such deep and hotly contested issues that even the choice of the starting point for discussion may seem to prejudge the outcome. In the present chapter, we try to take a 'naturalistic' approach, talking about the contents of scientific knowledge, especially in the context of *discovery*, in terms that scientists themselves might use in describing their work. In the following chapter, this 'folk' description will be complemented by a somewhat more general 'philosophical' analysis, indicating the unresolved issues that arise in the context of *justification*. We thus expose a variety of opinions concerning the nature of scientific knowledge, its ultimate validity, and its relationship to other bodies of organized knowledge. This discussion should help in deciding whether science can be distinguished from its competitors by any straightforward intellectual principle, and thus throw light on the internal sociological features of science (chapter 8), on the vexed question of the 'scientificity' of the social sciences themselves (§16.4) and on the place of scientific knowledge in our culture (§16.5). To keep the present discussion within bounds, we concentrate on the conventional natural sciences in their 'basic' mode, temporarily neglecting all possibilities of technological application. We also largely ignore the historical dimension, which brings the philosophy of science to life – at the expense of immensely complicating our understanding of it.

2.2 Description

Primary scientific information is essentially *descriptive*. According to the conventional metaphor, scientists 'explore the natural world', and endeavour to describe it 'as it is'. In such disciplines as geology, botany and zoology, there is a long tradition of simply recording what can be seen with one's own eyes, without interfering with natural processes. Although *natural history* is often belittled by scientists and philosophers, it is the source of a great deal of the elementary scientific knowledge on which more elaborate edifices of experiment and theory are actually built. The fact that much of this knowledge was recorded and re-recorded a long time ago does not make it less fundamental in principle. For example, the possible fit between the coastlines of South America and Africa, one of the keys to the modern theory of plate tectonics, was long evident from the scientific reports of navigators, as 'described' in world maps of several centuries ago.

But a *scientific* description of an object or event is not just any account of something seen or experienced. It is expected to conform to certain canons of 'accuracy', 'completeness', 'reliability', and so on. Although the criteria for such qualities are difficult to formulate explicitly, they are established by convention in any recognized scientific discipline. They define the characteristics of what we come to call the *facts* of that branch of science.

For the moment, let us not question the fundamental epistemological status of such 'facts'. At this unsophisticated level of analysis, a 'scientific' report such as 'the white swan sat for seven days on three eggs' may be taken to be no more, or less, 'factual' than a statement that 'the cat sat on the mat'. The point to be made is that it is no trivial enterprise to establish a body of scientific information conforming to these canons. This information can only be acquired by very deliberate action, that is, by *observation*. Thus, the facts of the science of ornithology do not derive so much from innumerable occasional accounts of 'bird watching' as from the patient and systematic observation of the behaviour of birds in all manner of environments and circumstances. Many major scientific disciplines, such as palaeontology or human anatomy, would make little progress without the special skills of observers who have been trained to scrutinize objects carefully, to notice significant details, and to record them accurately. Expert observation is not only one of the arts of scientific investigation: from the point of view of the practising scientist it is a major constituent of the scientific 'method'.

2.3 Generality

A description of the natural world in terms of *particular* facts would be quite unmanageable and quite useless. The essence of scientific knowledge is that it goes beyond individual items of information and encompasses them in *general* statements. The particular fact that 'this stone is rather like this bird's egg' may be perfectly true, but it is of negligible scientific interest compared with, say, 'the eggs of shore birds look very like pebbles'. The facts obtained by observation and other research are necessarily specific to the time and place where they were determined, but they have to be treated as in some sense *representative* of all that might similarly be reported in similar circumstances.

Once again, this sort of remark begs enormous questions. The significance of the names we use in referring to whole categories of individual items is one of the oldest issues of philosophy, applying in principle to all uses of language. In the theory and practice of scientific research this is more than an issue of principle: it is often a controversial *methodological* question with a direct influence on the contents of scientific knowledge.

In science we want to make general statements that are, so to speak, no less 'factual' than their individual elements. A great deal of scientific effort is therefore devoted to the task of discovering (or devising?) well-defined natural *classes* of objects and events. The essential question is: by what criteria is it advisable to treat certain individuals as practically *equivalent* to one another, and at the same time as *different* from another class of individuals? What are we letting ourselves in for when we refer to 'all *swans*', say in contradistinction from, say, 'all *geese*'? Is it a fact that these two classes of birds can be distinguished unequivocally, by reference to simple standard criteria?

The prime example of this type of scientific work is the discipline of biological *taxonomy*. Research devoted to the identification, description and classification of living organisms, according to their observable characteristics, is one of the oldest branches of science. Although it is often regarded as a relatively mundane activity, it is still the essential basis for all serious biological research, however sophisticated. The same applies to the identification and classification of mineral specimens, which is one of the factual foundations of geology. Much of this research seems very practical and empirical, and yet it can give rise to controversies which reveal deep divisions concerning the 'method' of science itself. Present-day debates over 'cladistics', for example, are rooted in philosophical questions which even have connections with conflicting political ideologies.

2.4 Patterns of fact

The differentiation of factual information into distinct categories is only the first step in building up an observational description of the natural world. An amateur naturalist identifying flowering plants in the fond hope of finding a rare new species might well be content with a 'Guide to Wild Flowers' where they are listed alphabetically by their botanical names. The data concerning comets are listed according to the year of their discovery. But a vast list of facts arranged under a succession of different headings is very far from constituting a body of scientific knowledge.

To give a general description of particular facts, one must somehow encompass them in a classificatory *scheme*. In biological taxonomy, for example, a structure of classes is exhibited: each *genus* is a set of *species*; each *order* is a set of *genera*; and so on, up a hierarchy of 'classes of classes'. In principle, any biological organism can be assigned to a unique category that has a recognizable place in this scheme.

Logically speaking, the construction of a more structured scheme of classification calls for little more than a broadening of the criteria by which individual items are assigned to primary classes. Statements such as 'This animal is grey, and that animal is tabby, but they have so many features in common that both are *cats*' are readily generalized to 'These cats are small, and these tigers are large, but these species have so many features in common that both belong to the genus *Felis*'. Philosophical puzzles start at the lowest level of classification, and do not become more serious in proportion to the complexity of the overall scheme.

In practice, however, as a classification scheme is extended to cover more and more members and classes of members, increasing attention must be given to the *ordering principle* by which it is organized. Rules have to be formulated concerning the criteria by which the various classes are to be distinguished. Thus, for example, it is rather obvious that cats and tigers are very similar creatures – but what is the rationale for considering all animals with backbones to be members of a single very large group, even though this is subdivided into innumerable species as different from one another as elephants and minnows? Questions like this, which cannot be answered by

2.4 Patterns of fact

reference to a limited number of direct observations, arise inevitably in the endeavour to give a general description of the natural world.

Any ordering principle that is consistent with the elementary facts that it classifies is itself, so to speak, a 'scientific fact'. But this leaves the way open to an infinity of 'artificial' schemes that would be entirely uninteresting. There would not be much point, for example, in classifying birds according to the predominant colour of their plumage, lumping blackbirds with crows and swans with seagulls. An ordering principle is only of scientific value if it generates a taxonomic scheme which is in some sense 'natural' to the aspect of the world to which it is supposedly applicable.

But what is the difference between an 'artificial' and a 'natural' scheme of classification? This, of course, simply begs the whole question of the validity of scientific knowledge, whether in the field of natural history or in any other scientific discipline. It also has the effect of directing attention to the characteristics of the assembly of observed 'facts' as a whole, and to the structural relationships between its subdivisions. In other words, it emphasizes the scientific objective of discovering *patterns* in the natural world, transcending all the details of primary observation. In any particular field of research there is no saying in advance whether this goal can be achieved: but if an 'interesting' pattern is actually perceived amongst the innumerable facts of nature, then this counts as an important general 'fact' in its own right. Thus, for example, one of the most significant characteristics of biological organisms is that they can be classified in a single tree-like pattern, where branches that have once separated never recombine – that is, despite superficial similarities, there are no hybrids between sharks and sealions, nor between birds and bats. This is just as much a 'fact' of biology as that birds have feathers and sharks have sharp teeth.

Biological taxonomy is the most obvious example of classificatory generalization and pattern perception in science, but the same basic processes are to be found in all fields of research. Indeed the 'typology' of a science – the method of classification that it applies to elementary observable 'facts' – characterizes that science. Thus, a piece of rock with magnetic properties might be regarded 'geologically' as *metamorphosed basalt*, 'chemically' as containing a particular *oxide of iron*, and 'physically' as an assembly of *ferrimagnetic crystals*. In each of these different sciences it would come under a different scheme of classification, related to a different type of ordering principle and hence having a place in a different type of 'pattern'. That is to say, the different sciences have traditionally been concerned with different observable aspects of the natural world, in which they have endeavoured to discover different types of abstract structure. Chemistry, we might say, is not about distinct objects or organisms, but about their material constituents, classified primarily according to the compounds they contain. These, in turn, can be arranged according to their elementary constituents, their molecular structures, etc., in a vast 'pattern' derived from the 'ordering principle' of the atomic hypothesis. Similarly, to modern physics the primary 'facts' of nature are 'events' such as the interactions of elementary particles, which can be ordered in time and space in 'patterns' derived from strict mathematical principles. There is a vast practical difference, of course, between noting that geese are not quite the same as swans, and discovering that electron neutrinos are not identical with muon neutrinos, but in their respective scientific contexts these can be considered observable 'facts' that have to be reckoned with in any general description of the world.

2.5 Investigation

Scientific work is normally much more active and purposive than the passive word 'observation' suggests. The goal of science is not the accumulation of factual information as such: it is to acquire knowledge in the form of significant general patterns of fact. Out of the plethora of potentially observable 'scientifically accurate' facts, very few are distinct and simple enough to be fitted directly into a well-structured classification scheme. Scientific research is therefore directed towards acquiring the special type of information that is likely to contribute to this endeavour.

In the development of any scientific discipline there are, of course, phases of *exploration*, where completely new information becomes accessible for the first time, and recorded unselectively with little regard for its ultimate scientific value. Yet even in a relatively descriptive science, such as human anatomy or field geology, the observations that are recorded are chosen because they can be classified according to an established ordering principle, and are therefore likely to contribute to the discovery of new 'patterns' of the same general type.

The most effective strategy of research, however, is purposeful *investigation* – that is to say, deliberate study of the circumstances that are thought to relate to an existing fact or idea. This usually takes the form of a positive policy of formulating specific *questions*, and then seeking the information needed to answer them. Indeed, this policy is so characteristic of scientific work that individual scientists are often supposed to have a peculiarly 'questioning' attitude of mind which makes them personally sceptical even in non-scientific matters (cf. §15.2).

And yet, as everybody knows, many important scientific discoveries have been made as if by chance. In the course of a scientific investigation – or even in the course of entirely unscientific activity – an observation may be made that seems somehow relevant to a question that was not consciously in mind at that time. Just occasionally, further investigation then leads to a significant scientific advance. *Serendipity* is such a familiar phenomenon in the context of discovery that it cannot be ignored, even though it defies formal analysis. The most that one can say is that experienced research scientists are acquainted with scientific knowledge on a diversity of topics, and are always alert to any apparent inconsistency between what occurs in the world about them and what might have been expected to occur in the circumstances. This is obviously connected with individual personal traits such as *curiosity*: there is really no way of separating the 'philosophical' dimension of science from the psychological aspects of scientific work (§15.1).

2.6 Instrumentation

Scientists would not be human if they did not make use of *instruments*. The anatomist's scalpel and the geologist's hammer, the microscope and the telescope, early became indispensable tools of scientific investigation. Whole scientific disciplines such as microbiology and astronomy have only been made possible by the development of devices that extend human perception into otherwise inaccessible domains.

This development seems so natural that it is difficult to see any distinction in principle between, say, observing the shape of the neck of a swan through a pair of field glasses and making the same observation, with greater difficulty, by the naked eye. Yet every scientific instrument, however simple, is susceptible to the criticism originally offered to Galileo's telescope – that the strange objects which he claimed to be seeing through it were not really up there in the heavens but were merely artefacts of the instrument itself. The characteristics of the instruments used in research cannot be dissociated from the 'facts' observed with those instruments.

The notion of 'observing' expands by analogy with the invention of more and more elaborate '-scopes'. A conventional compound microscope or reflecting telescope produces its visible image by the manipulation of ordinary light. What about the *electron* microscope, or the *infra-red* telescope, which make visual representations of patterns of electrical signals that could never be detected directly by the human eye? Is it proper to talk of 'observing' (perhaps even of 'seeing') a tumour in a section of a living brain by means of a computerized tomographic X-ray scanner, which produces its 'pictures' by calculation without even cutting open the skull? There seems to be no discontinuity, in practice or in principle, between Sherlock Holmes scrutinizing a suspicious footprint through his hand lens and Sir Martin Ryle investigating a distant galaxy through his radiotelescope!

Human perception is thus extended by instrumentation, not only in sensitivity but also in modality. The term 'observation' is expanded metaphorically far beyond the generation of images for direct visual inspection and spatial interpretation. In the physical sciences it has long been applied to the use of any device whose output represents any desired aspect of nature, by any symbolic means such as a 'pointer reading', a graph, a spectrum, or a number on a dial. Such devices, often of formidable complexity, are now in routine use in all scientific disciplines, supplementing or even superseding the traditional techniques of direct human observation.

Scientific instruments have the immense virtue of being free from *observer bias*. The *subjectivity* of eye-witness evidence is notorious, not only in the police court but also in more specialized tasks such as timing the transit of a star across a wire in the eyepiece of a telescope. A mechanical device cannot be influenced by emotional or other personal factors which often interfere with the unambiguous description of natural objects and phenomena. Of course any instrument has its own characteristic defects, which may introduce random or systematic errors into its symbolic output; but such defects can often be reduced to negligible proportions by deliberate redesign – an option that is not available when dealing with human observers.

The invention of more and more sensitive and accurate scientific instruments is thus a major component of scientific work (§11.2). This is not merely a technological factor, for the use of instruments is inextricably woven into the research process. A powerful research tool such as a mass spectrometer is not a passive observational device like a telescope. It must *operate* on the natural world (by taking a fragment of a specimen, vapourizing it, passing the gaseous products through electric and magnetic fields, etc.) and *transform* the consequential events into a symbolic output (i.e. electrical signals at the detector are coded into points on a chart). Because the device has been designed to isolate and analyse some particular feature of the world, this output – e.g. that this specimen contains one part in a billion of dioxin – is usually interpreted by scientists as a 'fact' on a par with the more familiar 'facts' of the everyday world – e.g. that this room contains six chairs.

2.7 Measurement

Instruments are often used scientifically to make *measurements*. That is to say, they present the results of their operations in *numerical* form. This is a recent development. It is practically impossible to describe a natural object without using the language of integers. The botanist counts the petals on a flower, because he has learnt that this is a significant classificatory 'fact'. Simple continuous properties, such as height or weight, are represented in terms of conventional standard units. Measuring instruments have become more and more sophisticated, so that it is now possible to measure a vast range of properties to an incredible degree of precision – but the use of a mass spectrometer to 'weigh' a molecule is not more profound in philosophical principle than the use that Archimedes made of a balance to weigh a gold coin.

The transition from integral measures (numbers of petals) to 'continuous' quantities (weights of coins) is, of course, far from trivial, both practically and logically. Within any scientific discipline, one of the major objectives is to discover what relevant aspects of nature can be described quantitatively, and how best to do this. The science of physics, in particular, is dominated by the theory of measurement, elaborating into dimensional analysis and the paradoxes of relativity and quantum mechanics. This is not to say that all 'qualitative' classification schemes are scientifically inferior to 'quantitative' ones (or, as is often asserted, that the methodology of physics should eventually supersede the methodologies of all other scientific disciplines) (§16.4); it just means that an observational account in numerical terms is extraordinarily effective as a means of discovering significant 'patterns' in the natural world.

An instrumental measurement is a very powerful type of observational operation. A numerical datum is a highly selective and abstracted representation of a phenomenon or an object. '5.678 grams' is a poor substitute for a real gold coin, but it carries with it a vast amount of classificatory information that is impossible to convey in 'qualitative' language – for example that the coin in question weighs more than a sheet of airmail paper, and less than a hen's egg, and so on, *ad infinitum*. Indeed, it implies the existence of certain types of classificatory scheme, such as a linear ordering according to magnitude, which cannot be ignored in the search for a meaningful general 'pattern' in the 'facts'. Thus, for example, if the eggs of ducks, geese and swans were found to weigh 30 grams, 50 grams and 70 grams, respectively, it would be perverse not to place geese 'between' ducks and swans in respect to that particular property. Again, if two distinct properties, such as weight and volume, have been measured, it is obvious that the objects in question should be classified according to the standard mathematical representation of number pairs – i.e. as a 'graph' or a two-dimensional 'map'.

In other words, by choosing to represent the natural world in numerical terms, we bring our descriptions into the sphere of *mathematics*. Once again, we need not go into the fascinating question of whether mathematics is entirely independent of the observational sciences. The point here is that if we make our measurements appropriately – e.g. if we can find a satisfactory empirical meaning for various abstract mathematical relations such as equivalence and addition – then we are permitted to manipulate the data mathematically and generate a great many new 'facts' that were not originally 'observed' directly. For example, *mass* and *volume* are both quantities that satisfy the elementary axioms of arithmetic. By dividing the mass of each egg (say), by its volume, we obtain a new *derived* quantity – the *density* – which turns out to be nearly the same for all the eggs in question. Thus we have discovered a very elementary classification 'pattern' for these particular 'facts'. This is certainly not the only way that scientific generalizations can be discovered and expressed in logical form, but it illustrates a whole scientific procedure, typical of the physical sciences, of setting up measurement operations whose results satisfy appropriate mathematical axioms and of exploring their arithmetical, geometrical or topological implications.

2.8 Experiment

Scientific investigation is not limited to the study of *natural phenomena* – that is, to the observation of events that occur spontaneously and are for some reason considered striking. Modern science is largely founded on the results of *experiments*, where the natural world is deliberately interfered with in order to observe the consequences. For some disciplines, such as astronomy or geology, whose significant events are remote and inaccessible in space or time, this method of investigation is not practicable. In general, however, the *experimental method* is almost synonymous with the practice of research. A science such as chemistry may be largely built around a characteristic experimental *methodology*, such as the technique of bringing known compounds together under carefully controlled conditions and observing the reactions that occur.

The notion of an experiment covers such an enormous range of scientific work that it eludes precise definition; but considered as a general metascientific method of investigating nature and making discoveries it is usually supposed to have certain characteristic features.

First of all, an experiment is *empirical*. It is performed in the real world, in real time, on real objects, and gives factual results. Indeed, from this point of view, there is no distinction in principle between 'observational' and 'experimental' information, whether obtained by direct human perception or by means of instruments.

But an experiment differs from a mere observation in that it is *contrived*. The events to be studied are made to occur as far as possible under carefully controlled circumstances, which are often extraordinarily artificial and unnatural. This is particularly the case when elaborate instrumental techniques are employed in the form of experimental *apparatus*. It has been said that God Himself does not know what will happen in certain experiments using high energy particle accelerators because these events have never occurred in the Universe before!

The artificiality of an experiment shows that it must be *intentional*. The effort that has to go into devising it and making it 'work' is never entirely pointless or playful: it has a rational purpose, such as observing a phenomenon, or exploring an unknown domain. This rationality applies, of course, to all the surrounding circumstances of the experimental situation, such as uniformity of conditions, protection from interference and measurement of significant results (§3.2).

Finally, a true experiment must be, in some sense, *original*. In an ideal form it is supposed to generate new information through a novel 'experience' not previously reported to science. Many experiments are, of course, designed to *replicate* previous experiments, and to check that their results are *reproducible* (§3.2), but this only implies that these results are not yet considered quite certain, and might turn out different and new. Even though practical science education is largely based on standard laboratory 'experiments', the notion does not properly apply to the mere repetition of an operation whose outcome is not in doubt.

2.9 Scientific laws

As anyone who has tried to make a subject index knows, any general classification scheme tends to ramify endlessly. All the results of scientific observation and experiment would eventually be rendered useless by their sheer multiplicity and diversity if they could not be reduced to intelligible form by *simplifying* generalizations. Research is therefore directed towards the discovery of 'patterns' of classification whose structural principles can be stated very succinctly, even though they cover a multitude of particular instances.

The simplest possible 'regularity' that might be noted in an assembly of 'facts' is that of *invariant association*. A general statement such as 'all swans have feathers' or 'all electrons have spin' has the effect of conflating two observational categories. It is no longer necessary to determine whether or not it has feathers if the creature is known to be a swan, or to make a separate measurement of its spin if the particle is known to be an electron. The amount of detail required in describing the world is thus significantly reduced.

Of course most regularities of this kind are not very profound. Indeed, it may be difficult to assert them independently of the classification criteria of their components. It may be, for example, that a 'swan' has been *defined* as a creature with (amongst other characteristics) 'feathers', or that the notion of a 'particle with spin' has been exemplified solely by measurements on electrons. Nevertheless, any such invariant association of characteristics, if it is indeed a 'fact', helps to reduce the numbers and dimensions of the categories that may be needed in constructing the 'scheme of things'.

When such a regularity is considered to be highly significant for science, it is often called a *law* of nature. Scientific laws are formulated in the course of scientific work in many different ways. They may codify the results of laborious and lengthy observations, as in the case of the arithmetical laws for the succession of eclipses discovered over many centuries by Babylonian astronomers, or they may represent the outcome of some imaginatively contrived experiment, as in the case of Boyle's Law describing the relationship between the volume of a gas and its pressure.

In scientific practice, a law usually denotes a definite association between *empirical* characteristics of natural phenomena – i.e. of features that are in some broad sense 'observable' – rather than a relationship between abstract 'theoretical concepts'. But there is nothing clear about this distinction. Some laws are quite clearly *phenomenological*: for example, Mendeleev's Periodic Law of the Elements summarized an immense amount of chemical knowledge, but seemed at the time to be quite unconnected with any other general scientific principle. On the other hand, more 'fundamental' laws such as Newton's laws of motion and the various laws of thermodynamics apply to 'facts' that are very far from being directly observable. From a philosophical point of view, the notion of a 'law-like' statement also extends to elaborate associations of instrumental measurements that can only be expressed compactly in mathematical symbols – for example, the laws of black-body radiation, which were the starting point for quantum theory.

The question of the validity and epistemological status of scientific laws is one of the principal focal points of the philosophy of science. It is very difficult to talk about them at all without appearing to take a particular standpoint on this question. The connection between a 'law of nature' and a 'law of the state' is now, of course, just a verbal remnant of an outdated metaphor. But should one talk of a scientific law being 'discovered' or 'constructed'? Is there any sense in which some laws are more 'fundamental' than others? Are the regularities described by scientific laws produced by 'chance' or are they somehow 'essential'? These are questions that must be deferred until the next chapter (§3.9) for they cannot be separated from opinions about the status of scientific knowledge as a whole.

2.10 Explanation

It is universally agreed that one of the major goals of science is to *explain* the facts of nature and the laws that seem to govern them. What does that mean? 'Explanation' is an extremely deep notion, far beyond precise definition. But the characteristic form of a scientific explanation is a rational argument linking an assembly of empirical facts with a general conceptual scheme. It is, so to speak, one of the types of relationships between 'factual' and 'theoretical' scientific statements.

Thus, any well-formulated scientific law may be counted to a modest degree as a step towards explaining the observation that it sets in order. It is natural to say, for example, that sodium interacts with water to form an alkali *because* it is an element in Group I of the Periodic Table, and according to Mendeleev's Law *all* elements in Group I have this property. This would be even more 'explanatory' if the element in question were newly discovered, and this particular property had not yet been checked first. A scientific law asserts a generalization which to some extent 'explains' any further instances of itself.

A much more convincing type of scientific explanation is one that links a generalized class of facts with an intellectual structure derived from a different empirical domain. Thus, for example, there is immense satisfaction in explaining the chemical phenomenology of Mendeleev's Law by physical representation of an atom as a charged nucleus surrounded by electrons. An essential feature of the notion of an explanation may be that the *explanans* must be somehow 'more general' than the *explanandum* it is supposed to explain, even if the latter is itself a general classification principle or scientific law. This is why strong explanatory theories often refer to 'invisible' entities (e.g. 'atoms', 'electrons', 'genes') since they must be constructed of less particular and more abstract components than can be grasped by direct observation.

Ideally, the explanatory relationship should be strictly logical. There are many cases, especially in the physical sciences, where the explanation is a straightforward deduction of a special case from a general *covering law*. For example, Kepler's 'Laws of Planetary Motion' can be derived by a simple mathematical calculation from Newton's laws of motion and the Principle of Universal Gravitation. This sort of mathematical deduction is not really as logically watertight as students are led to believe, but it satisfies as high a standard of formal rigour as is ever achieved in science in practice.

Scientists usually have to make do with explanatory arguments that are far from logically compelling. In extreme cases, a mere analogy may be considered a sufficient connection to count as an 'explanation'. Darwin, for example, explained the immense diversity of existing and past biological species by analogy with the diversity of breeds of domestic animals produced by artificial selection. This metaphorical parallel between 'artificial' and 'natural' selection is still accepted even though it has not yet been 'proved' by a strict mathematical analysis. Each scientific generation sets itself higher and higher standards of explanatory rigour, and yet it usually turns out that what passes for a convincing scientific explanation is still a matter of opinion, justified by reference to undefinable criteria such as 'simplicity', 'generality' and 'fundamentals'.

2.11 Cause and effect

A traditional form of scientific explanation is to argue that the *cause* of a certain event B was the previous occurrence of a specific event A. Thus, for example, the death of this child was caused by its having contracted diphtheria, which was caused by infection with a specific bacterium – and so on. The notion of a *causal chain* comes from the world of everyday life: every child knows the tale that starts with the loss of a horseshoe nail and ends with the loss of a kingdom. This notion is readily generalized scientifically from particular events to universal categories, and often appears in the statements of scientific laws. It was a great step forward in medicine to assert as an observational regularity that a great many diseases were evidently caused by bacterial infection.

But modern scientific discourse does not make great use of the terminology of cause and effect. At an early stage in the exploration of a new field it is revealing to notice that certain phenomena invariably occur together in the same order – that the symptoms of diphtheria always follow upon infection with *Corynebacterium diphtheriae* – and then to go on to test this regularity by deliberate observation and experiment. But mere invariant association in close temporal order does not constitute a *causal* relation. We do not say, for example, that the radio time signal for 6 o'clock 'causes' the news broadcast that follows it, or is 'caused by' the weather forecast that precedes it, even though this chain of events is sufficiently regular to count as a sociological generalization. The arguments surrounding the assertion that cigarette smoking 'causes' lung cancer illustrate the difficulty of constructing a satisfactory scientific explanation out of a mere statistical correlation in the absence of more direct laboratory evidence for a causal relationship.

A causal explanation thus calls for much more information than mere association of event A with subsequent event B. It is essential, for example, to check that the surrounding circumstances are sufficiently alike, in each case of the association, to allow the 'effect' to occur. To say that B follows on A provided that 'other things are equal' seems innocent enough, but the question what 'other things' might be relevant to these events can only be answered by reference to a whole range of general principles and theories that are not actually stated in the causal connection. How can we account for the fact that many non-smokers also die of lung cancer, except by reference to medical knowledge about other carcinogens, about the spread of secondary cancers in the body, and, ultimately, to the basic principles of anatomy, physiology, pathology and biochemistry? Thus, a causal argument, like any other type of scientific explanation, is only satisfactory if it is embedded in a larger, more general conceptual scheme.

The notion of cause and effect becomes interesting when human volition is involved. Sometimes we know enough about a particular system to be fairly certain that if we set it in motion, by action A, then a desired event, B, will eventually follow. It may be a gross oversimplification to say that detonating the fuse of a nuclear warhead is the *scientific* cause of the ensuing fission reaction – and the eventual destruction of a city – but this is an example of the deliberate contrivance of a causal chain directed towards the achievement of a particular 'effect'. Such chains are thus highly significant in the application of scientific knowledge, even when they are linked by invariant associations that are 'inexplicable' in a more general sense.

2.12 Models

The results of a scientific investigation are usually much too numerous and diverse to fall obligingly into obvious classificatory patterns, or to follow one another in simple causal chains. An elaborate formal scheme of interacting 'laws' might be set up to express all the regularities and 'anomalies' that had been observed, but this would soon become much too complicated to count as a satisfactory 'explanation' of the facts in question. When these no longer seem to generate their own ordering principle, guidance must be sought elsewhere. The most fruitful source of explanatory schemes in science is *analogy*. It is often possible to give a satisfactory account of a body of scientific facts by reference to a *model*.

The notion of a scientific model is very broad. It might be defined as a real or imagined system whose structure is similar in important respects to the system under investigation, but this does not get us much further. To what extent, for example, need the model be 'similar' to the system being studied. The scale models used to investigate the sea-going characteristics of ships are obviously similar in shape, mass distribution, watertightness, etc. The molecular models used by chemists are enormously scaled up, and quite unlike their originals in almost every material characteristic except the relative spatial arrangement of their components. The Bohr model of an atom as a miniature Solar System still has valuable explanatory power, even though the 'planets' represent particles which cannot even be localized precisely. Even in scientific thought the boundary between 'analogy' and 'metaphor' cannot be strictly defined.

Nevertheless a model can often provide an 'explanation' for a vast range of observed facts. At the beginning of the nineteenth century, for example, the atomic model provided a simple rationale for an immense amount of empirical knowledge about chemical compounds and their reactions, including a great deal of numerical detail concerning combining weights even in reactions that had not previously been studied. It was further extended to cover numerous other properties of chemical compounds, such as their crystalline form, which could easily be explained in terms of the geometrical arrangement of spherical objects in space – right down to the elementary calculation of sizes and distances that were crucial in determining the structure of DNA, and hence explaining another vast body of biological phenomena. In this case, the explanatory scheme is so realistic that we may say that we have discovered the *mechanism* of genetics.

In the physical sciences, models are often so well defined that their behaviour can be analysed mathematically. Thus, for example, the analogy between the propagation of light and the propagation of waves on the surface of a pond may be taken sufficiently seriously to suggest the calculation of various diffraction phenomena that might be observable in both systems. Indeed, there soon comes a point in such a calculation where the material nature of the model itself is forgotten. The general equations for wave propagation, whether for light, or surface waves, or sound waves, or whatever, then constitute a coherent explanatory scheme or 'model' for all the phenomena in question. It is customary nowadays, for example, to refer to a computer 'model' of the atmosphere, even though this consists of nothing more than a programme for manipulating observed measurements of temperature, pressure, humidity, etc., according to the dynamical equations of meteorology. The notion of a model thus extends into a purely symbolic domain, where there is only an abstract similarity between the original system and its model.

2.13 Theory

This brief account of the cognitive goals of science began in the realm of factual description. We have now reached the opposite pole – the realm of *theory*. Approached from this direction, scientific theories appear as ordering principles that explain general classes of observational and experimental facts, including the taxonomies, 'laws', causal chains and other empirical regularities that are discovered about such facts. A well-founded theory, covering a wide range of facts to a high degree of accuracy, is thus the most compact and manageable form in which scientific information can be recorded, manipulated, used or understood. It is the vehicle by which a description of natural phenomena is expressed as scientific *knowledge*.

But theory has standing in its own right. We have already noted that an explanation must be more general and abstract than its *explanandum*, so that explanatory schemes nest inside one another, expanding in a hierarchy of increasing coverage. At a certain stage, as we move away from simple 'laws' and other phenomenologies, we tend to lose sight of the empirical 'factual' characteristics of the entities under discussion, and treat them purely as *concepts*, existing only in the domain of thought. Theories belong unequivocally to the world of ideas, and can only be expressed or communicated in symbolic form, such as by words, mathematical formulae, or diagrams. They assert structural relationships between concepts, which can be further manipulated in abstraction, according to logic or other 'laws of thought'. *Theorizing* thus becomes a distinct activity within science, temporarily disconnected from the natural world and not immediately directed towards the explanation of observed phenomena.

There is no saying what will make a satisfactory theory in a particular field of research, but certain broad characteristics are essential. In the first place, a scientific theory must be *rational*. It must hang together logically, without obvious inner contradictions; otherwise it could not be related unambiguously to experience (§3.7). In practice, this is quite a strong condition, since it is not always easy to determine the self-consistency of a collection of closely coupled formal propositions or mathematical equations. One of the advantages of constructing a theoretical system around a material model is the knowledge that it can in fact be 'realized' without self-contradiction. For example, the Mendelian theory of combining and mutating 'genes' was strengthened by the discoveries of molecular biology, showing how these abstract entities could be 'modelled' chemically. But there is no absolute necessity to follow contemporary metascientific fashion, which tends to equate 'theories' with 'models', presumably on the grounds that only well-articulated theoretical structures that can be realized as coherent systems are worthy of scientific consideration.

Another essential property of a scientific theory is that it should be *relevant*. A beautifully articulated and self-consistent structure of abstract entities is of no scientific interest unless it is accompanied by *interpretative* principles relating it to the empirical world. Thus, for example, Einstein's General Theory of Relativity would be no more than a fascinating exercise in pure mathematics if it did not tell us that the symbols x, t, E, etc., stood for measurable quantities such as position, time, energy, and so on. Even though these principles may not play a significant part in the internal theoretical manipulations, they are essential elements in the theory as a whole.

Finally, if a theory is to be of scientific use, it must be *extensible*: it should 'explain' more facts than it was originally intended to cover. To give a famous example: Maxwell produced a set of theoretical equations that summarized the known facts about electricity and magnetism. This would have been no more than an elegant formalism if he had not shown, by an imaginative change of one of the terms and some straightforward analysis, that these equations had solutions in the form of waves. A theory intended to cover electromagnetism was thus extended to explain the wave properties of visible light. This characteristic of theories is not only highly desirable: it may also be one of the characteristics that differentiate scientific theories from other conceptual structures.

2.14 Hypotheses

How can theories be 'discovered'? Unlike observational or experimental facts, they cannot be obtained by deliberate investigation. They do not spring automatically to the eye like the hidden patterns in a test for colour blindness! However convincing and 'real' they may seem in retrospect (§3.9), scientific theories are *mental* entities that have to be constructed by human thought. In the absence of a complete scientific or metascientific account of 'creativity', this stage in the 'method' of science escapes rational analysis. A theory comes into being, in the mind of a particular scientist, in the particular historical circumstances of his or her research. There is much to be learnt from the records of such episodes in the history of science – for example, in the notebooks of a highly creative researcher such as Michael Faraday. There is also much to be said about the technical and social context in which any such episode may have occurred (§7.3). But there is always an imponderable element of chance and/or personality which is not explained away by being called 'imagination' or 'intuition'.

The crucial point for a theory is just the moment of creation, before it has been subjected to the more methodical processes of substantiation. At this moment it can be no more than a *hypothesis*, formulated mentally as a candidate for further study, worth consideration for its explanatory potentialities, but with no firm commitment as to its eventual validity. It may even present itself simply as *conjecture*, suggesting a possible conceptual relationship without immediate concern for consistency with other theoretical principles or empirical facts. The sequence from 'conjecture' through 'hypothesis' to 'theory' suggests an increasing degree of coherence and scientific certainty, although it must be admitted that scientists use these terms loosely and rhetorically, and seldom have good grounds for differentiating them in practice.

But scientific hypotheses cannot be created out of nothing. Even the wildest conjecture must be expressed intelligibly. It must be communicable in a form that can be understood, and eventually connected with other conceptual relationships or with the results of observation. In most cases, this can be done within the conventional formalism of the science in question. For example, chemical hypotheses can usually be expressed diagrammatically in terms of three-dimensional atomic arrangements, whilst theoretical physics is almost always communicated in the symbolism of advanced mathematics. Indeed, every scientific specialty has an established repertoire of useful concepts and formalisms out of which new theories can normally be constructed (cf. §7.3). The vocabulary of scientific publications is not only notoriously esoteric: in any particular field of research, this vocabulary may actually be quite limited, as if most of the facts were being satisfactorily explained in terms of only a few standard theoretical concepts.

When the existing conceptual repertoire seems inadequate, a suitable theoretical structure may often be borrowed from some other specialty or discipline. Theory building by analogy transfers models and other conceptual entities from field to field of science. This process is so effective that it sometimes seems as if all theorizing must be by analogy, but this may be only another way of saying that theoretical concepts can only be communicated by verbal, mathematical or diagrammatic symbols. Indeed, it is difficult to disentangle this argument from the more general proposition that every use of language is essentially metaphorical.

Is there, perhaps, a limited overall repertoire of fundamental scientific concepts? Although there is nothing in the methodology or philosophy of science to impose such a limit, it is remarkable how, in the history of science, a small number of characteristic *themata* (to use Gerald Holton's term) have figured in successive theories. Fundamental physics, for example, can almost be expounded in terms of the antithesis between opposing themata, such as 'atom' against 'continuum', or 'symmetry' against 'disorder'. Although the notion of an intellectual *thema* is beyond strict logical analysis, it opens fruitful avenues for further metascientific study.

2.15 Problem-solving and the growth of knowledge

A novel scientific hypothesis is only the germ of a discovery. The next step is to subject it to further analysis to test how well it fits into what is known or might be found out (§3.7). Does it really explain all the facts for which it was originally conceived? Are there other known facts which it might explain, or with which it might be inconsistent? What are its empirical implications? Could an investigation be devised to observe these implications? To what extent is it logically consistent with other supposedly well-founded theoretical schemes?

In practice, hypotheses are seldom unique. At any given moment, several different theories may be competing as explanatory schemes for a particular body of facts. Which should be preferred? This choice need not depend solely on the degree to which they seem to fit these facts, and to be logically consistent with pre-existing theories. There is a definite advantage in choosing a 'working hypothesis' that is simple enough to be easily formulated, with a clearly articulated conceptual structure, whose theoretical and experimental implications can be unambiguously investigated – at least until it has been shown to be untenable. An 'economical' scientific hypothesis is not merely aesthetically pleasing: it has functional value as a source of further research action and inspiration.

'Theorizing' is not, therefore, a separable component of the research process. Although it may have its armchair phases, it is derived from and leads rapidly back to the outwardly more strenuous labour of observation, experiment and measurement. In everyday life, 'theory' and 'fact' are often regarded as polar opposites. In science, this is a creative opposition, to be seen in the dynamical interaction between the 'rational' and 'empirical' aspects of scientific work. In any particular field of research, this dialectic is usually extremely complex, and convoluted. Expert knowledge is needed to decide the status of the observational data, experimental projects, formal calculations and speculative hypotheses that are mobilized around new theoretical proposals, or that are called into question by an 'inexplicable' experimental result. The purpose of a specific experiment may be, indeed, to test a particular hypothesis; or the form of a specific hypothesis may have been dictated by the desire to explain a particular set of data, but this sort of analysis is seldom meaningful on a broader intellectual or practical scale.

For the individual participant, scientific research separates more naturally into a succession of *problems*. The scientist working in an established discipline does not face a virgin natural world awaiting exploration or exploitation. He or she is normally immersed in an intellectual, technical and social milieu in which certain 'questions' seem to be demanding answers. These questions are very diverse and often very ill-defined, corresponding to the width and depth of the problem situations where scientific methods may be applied. Some questions, arising from the technological applications of science (§10.1), are clearly *practical*: 'How is the fuel economy of a motor car engine to be improved?' or 'Does this drug cure cancer?' Others are directed solely to gaining an *understanding*: 'Where do comets come from?', or 'What is the mechanism of bodily growth?' Each of these questions can be fragmented into innumerable interlocking sub-problems which can be tackled more or less independently: 'How can the fuel–air mixture be made to mix homogeneously in the combustion chamber?', or 'What is the chemical structure of the growth hormone discovered just last year in Japan?' When scientists are asked what they are trying to do, they talk of solving relatively well-posed problems of this kind.

The elaborate social and intellectual machinery for thus dividing the labour of research will be studied in later chapters, but the very notion of a scientific problem really requires detailed philosophical explication. Although it must be capable of relatively distinct formulation, it cannot be like a crossword puzzle or mathematical exercise, known to have a unique solution. It arises out of the research process itself, as it goes on, and is both novel and open-ended, so that there is no general way of tackling it except by a sophisticated version of 'trial and error'. It is part of the craft of research to be aware of new problems, to formulate them in a way that seems to make their solution feasible, and to judge whether, eventually, an acceptable solution has been found. In some cases, for example, a few relevant facts or empirical regularities would count as significant progress; for other problems, nothing less would do than a thorough explanation of all the facts in terms of a well-established covering theory (§3.8).

Fortunately, this characterization of science as 'problem-solving' does not invalidate the more traditional discussion of the various empirical and rational aspects of the process of 'discovery'. The point is that the 'trial and error' method for solving scientific problems involves all these aspects of research. Phases of observation and experimentation interleave with theoretical speculation and the postulation of general laws. In the end, a solution is 'discovered', and shown to be satisfactory by empirical tests. Scientific work may not be the direct exploration, description and explanation of the natural world, as idealized in some naïve accounts of the discovery process. It is seldom easy to discern the application of a clear 'method' in the detailed technical history of a research project. And yet the conventional metascientific categories of observation, experiment, measurement, model building, theorizing, etc., provide the framework within which scientific knowledge actually grows.

As a result of this general process we supposedly arrive at good, sound, knowledge. But there's the rub! By what standards is our science 'good' or 'sound'? By what criteria can we assert that a scientific theory has been tested and found to be valid? What is it about the natural world that science really discovers? What do we mean by a scientific fact, and how reliable are scientific laws? Can a theory, born very tentatively as a mental construction, tell us what things are really like? It is impossible to continue this general discussion of the cognitive dimension of science without enlisting the terminology and elementary principles of the philosophical analysis of the meaning of science in the context of *justification*, which we enter in chapter 3.

Further reading for chapter 2

The main themes of the philosophy of science are so interwoven that the student should read the next chapter before turning to the general texts to be recommended there.

Many of the topics discussed in the present chapter are written about from the viewpoint of the working scientist in

> W. I. B. Beveridge, *The Art of Scientific Investigation* (first published 1950, reprinted by Vintage Books: New York).

A much more refined analysis of scientific discovery is given by

> M. Polanyi, *Personal Knowledge*. London: Routledge & Kegan Paul, 1958 (pp. 120–31):

this work is too large and deep to be read as a whole, but remains one of the major classics of science studies, and a prime source of insight from the point of view of an outstanding scientific mind.

'Scientific Inquiry: Problem Solving on Artificial Objects' is the title of an important chapter in

> J. R. Ravetz, *Scientific Knowledge and its Social Problems*. Oxford: Clarendon Press, 1971 (pp. 108–45).

The connection between scientific problem-solving and cognitive science is developed by

> M. de Mey, *The Cognitive Paradigm*. Dordrecht: D. Reidel, 1982 (pp. 202–26)

The notion of scientific themata was introduced by

> G. Holton. *Thematic Origins of Modern Thought*. Cambridge, Mass: Harvard University Press, 1973 (pp. 47–68)

Some of the features of serendipitous discovery are indicated by

> B. Barber & R. C. Fox, 'The Case of the Floppy-Eared Rabbits: An Instance of Serendipity Gained and Serendipity Lost', reprinted in *The Sociology of Science*, ed. B. Barber & W. Hirsch, pp. 525–38. New York: The Free Press, 1962

Details of various accounts of a famous serendipitous discovery are compared instructively by

> S. W. Woolgar, 'Writing an Intellectual History of Scientific Development: The Use of Discovery Accounts', reprinted in *Sociology of Scientific Knowledge: A Source Book*, ed. H. M. Collins, pp. 75–102. Bath: Bath University Press, 1982

3
Validity

'[The Scientist] must appear to the systematic epistemologist as a type of unscrupulous opportunist: he appears as a *realist* in so far as he seeks to describe a world independent of the acts of perception; an *idealist* in so far as he looks upon concepts and theories as free inventions of the human spirit (not logically derivable from what is empirically given); as *positivist* in so far as he considers his concepts and theories justified only to the extent to which they furnish a logical representation among sensory experiences. He may even appear as a *Platonist* or *Pythagorean* in so far as he considers the viewpoint of logical simplicity as an indispensable and effective tool in his research.'
                                                                    Albert Einstein

3.1 Epistemology

Scientific knowledge takes a variety of forms, from the most obvious descriptive facts to the most abstruse and speculative theories. Yet it is often treated as a single body of information, of peculiarly high credibility. The fundamental concern of *epistemology* is how much of this knowledge can be considered true, or how firmly it should be believed.

The history of science should dispel any notion that all science is true. There are innumerable cases of elementary errors of observation which were long held to be facts. Despite their best endeavours, scientists of previous eras 'discovered' and came to believe many ideas that we now consider to be quite mistaken. In all modesty, we must suppose that the science of our own time asserts the truth of some theories that are just as wrong as famous scientific errors of the past. Modern science must surely have its equivalents of the spontaneous generation of life, or the 'caloric' model of heat. Since scientific knowledge often contradicts itself, from generation to generation, it cannot all be true at all times. In practice, science is always prone to error and always open to correction.

But are these errors merely human imperfections in carrying out a procedure that is ideally capable of generating unquestionable truth? The research processes outlined in the previous chapter certainly do not add up to a mechanical technique for producing such a result. This applies particularly to the later phases in which conjectures and hypotheses (§2.14) are tested and validated. The credibility of a particular item of scientific theory clearly depends upon the extent to which it has been subjected to such tests and not found defective. Nobody supposes that a recent wild conjecture on, say, the origins of life on earth is true just because it has been put forward by an eminent molecular biologist: the central epistemological question is whether there is in principle a method by which a scientific theory can *eventually* be made *perfectly* certain.

The 'method' of science is much more amenable to philosophical analysis in the 'context of justification' than in the context of discovery. Whether or not these phases in the research process should be considered as different as, say, the work of the judge and the work of the detective in law enforcement, this is a convenient way of setting out the stages of thought and action leading to well-founded scientific knowledge.

3.2 Empiricism

For a start, the notion of a scientific 'fact' (§2.2) upon which so much of the credibility of science depends, calls for critical analysis. But at what level should such an analysis start? Nobody disputes that a great deal of what is reported as scientific information is 'factual' in the most ordinary sense of the word. A statement such as 'swans have feathers' or 'mercury is a liquid at room temperature' would seem to be as true and as certain as any statement could be about the everyday world.

This immediately suggests a standard of credibility which one might hope *all* scientific information could eventually attain. Whatever philosophers may doubt, most scientists and non-scientists would be entirely satisfied if all science could achieve the epistemological status of *empirical* truth. When scientists insist that science is 'nothing more than common sense, writ large', they are asserting that electrons and genes and black holes and pterodactyls are just as credible and real as most of what we feel sure of in the life-world, such as tables and chairs, cats and dogs, or uncles and aunts.

But even if this is accepted in principle, it does not look at all obvious in practice. Anybody who has tried to fathom out a very esoteric scientific theory, such as the immunity mechanisms of the body, or the quantum chromodynamics of quarks, needs a good deal of convincing on this point. It is thus essential to examine the steps by which 'common sense' is being extended far beyond its normal scope, into conceptual domains that might also include a quantity of uncommon nonsense.

The difficulty is that our belief in the truth of 'factual' statements about the everyday world then needs deeper analysis. But that would draw us into the traditional philosophical debates on solipsism and other variants of extreme scepticism. The philosophy of science is connected with questions such as whether the world exists when nobody is looking at it, because scientific empiricism seems the simplest and most lucid example of the way in which all our knowledge of the external world is generated. Although we shall not pursue this point here, scientific knowledge is often held to be an ideal model of human knowledge in general.

Factual scientific knowledge is certainly considered to be superior to ordinary common sense in several significant respects. In science, a deliberate effort is made to eliminate some of the known imperfections of human perception and observation. In everyday matters, the question 'What exactly are the facts?' is seldom posed rigorously, but when it is, a convincing answer is often difficult to obtain. Honest witnesses in a court of law frequently give contradictory testimony, even on the most elementary facts of time and place. The philosophy of science concentrates on the problem of justifying *theories*: the first task of the research worker is to justify the *facts* that he or she claims to have discovered. It is seldom easy to bring these up to the same standard of credibility as indubitable everyday realities. This is a peculiarly heavy responsibility when the investigation leads into unexplored domains, or yields results at variance with received opinion. Thus, for example, the early reports from Australia that duck-billed platypuses laid eggs were far more difficult to validate decisively than the corresponding statement about ducks, and the observation that meteorites actually fell out of the sky was laughed to scorn until overwhelming evidence had been obtained to justify it.

The practical methodologies of research are directed against two major sources of empirical uncertainty. The first such source is *subjectivity*. Although human perception is remarkably sensitive and discriminating, it is easily affected by bodily and mental factors that vary erratically from person to person. In science, therefore, the laboratory notebook over-rules the research assistant's memory, and a photograph takes precedence over a hand-drawn sketch. As we have already noted (§2.6) scientific *instruments* are used, not only to enlarge the range of the human senses but also because they detect, measure and record phenomena without personal bias.

Scientific knowledge often claims to be *objective*, as if free from all subjective influences. But this term could only be applied strictly to information that has been obtained without human intervention, such as the print-out from an automated scientific instrument. In reality, there is always a component of human judgement in the design of such an instrument, and in the interpretation of the data it produces. The best that scientific methodology can do is to try to neutralize subjective factors by playing off one human observer against another, and then only report what they all agree on. In the behavioural and social sciences (§16.4) this is a peculiarly difficult problem, but if one is not to beg the whole question of scientific epistemology one must not insist on a more rigorous standard of empirical 'objectivity' than *intersubjective* agreement on the facts (§8.6). An objective scientific fact should be, so to speak, like the Houses of Parliament; different people may see it from slightly different personal viewpoints but it is such a salient feature of the landscape that nobody seriously doubts that it exists independently of all observers.

The other major source of empirical uncertainty is *contingency*. Science has no use for unique objects or events which cannot be classified according to some rational principle and thus treated as representative of some general category (§2.3). The facts of interest to science must in some sense be *reproducible*. It is essential to be able to show that for any particular object or event there are others so like it in certain respects that they can be put into the same category and henceforth treated as scientifically equivalent. This is impossible if apparently significant empirical differences between individual specimens are being produced by chance.

For natural 'objects' such as biological organisms, this is an integral part of the fundamental principles of classification. Thus the formal taxonomic criteria for the identification of a species imply that one need not take into account other observable variations between individual members of that species: for example, research has shown that the number of lobes on each leaf of an oak is a significant classificatory feature, whereas the number of branches on the tree is not. But when new *phenomena* are being described, the circumstances under which they are said to occur may be extremely complex and must be precisely defined. There is always the possibility that the phenomenon has not been brought about by the recorded circumstances, but has been produced by an 'accidental' cause that was not observed or not reported. For example, when sharp regular pulses of radio waves were first observed from certain radio sources, months of research had to be done to eliminate other possible causes of this extraordinary phenomenon, such as intermittent electrical contacts in the apparatus, or the ticking of somebody's electric clock.

This is one of the reasons why empirical scientific work is often described generally as *experimental* (§2.8). Facts obtained by passive observation are less certain than the results of contrived experiments, which can be designed to minimize accidental influences on the course of events. Indeed, even a good experiment is not sufficient in itself. The scientific errors that can arise from uncontrolled variations in the external conditions can be so serious that it is usual to *replicate* an experiment before concluding that its results are truly reproducible. The repetition of experimental investigations, in slightly different forms, by independent groups of researchers (as a guard against observer bias, and other subjective factors) is a regular feature of the empirical methodology of science. This is what has to be done, in the context of justification, even for 'factual' scientific knowledge.

3.3 Phenomena and sense-data

The basic strategy of empiricism is to build up a body of strictly factual 'observations' whose validity can be made quite secure by procedures derived from everyday common-sense reality. Some philosophers have suggested that scientific statements framed in 'observational' language should be given a different epistemological status from those using 'theoretical' terms. Theories could then be constructed on these unimpeachable foundations, but could be regarded as hypothetically insecure, and potentially corrigible, without throwing doubt on the real truths that science has discovered.

Something like this certainly happens from time to time in the course of research. When a hypothesis is disconfirmed experimentally (§3.7), or found to be logically faulty, one must go back to the facts that it was supposed to explain and start again with some new hypotheses. But this is always a tactical move, representing a retreat to a position which is only relatively more 'empirical'. At any given moment in a given field of research there may be a fairly obvious distinction between well-established 'facts' and more conjectural 'theories', but this distinction is not precisely defined and seldom remains fixed over a period of time. When, for example, we measure the temperature in an experiment, we treat this observation as unquestionably empirical, quite forgetting the past research, both theoretical and experimental, that has gone into defining temperature as an 'observable quantity'.

It is now generally agreed that there is no such thing as a purely 'observational' language for normal scientific communication, and that ordinary empirical scientific observations are heavily 'theory-laden'. Scientific knowledge can never be quite self-contained. The method and circumstances under which an observation is made cannot be described accurately without reference to other scientific facts or concepts; for example, chemical reactions cannot be replicated without measurements of physical quantities such as temperature and pressure, which are certainly not elementary, everyday qualities. The instruments, such as telescopes, microscopes and cameras, used to extend human perception, or to make 'objective' observations, are seldom simple 'common-sense' devices. They are designed to work by virtue of various general scientific principles, such as the laws of optics, and have to be tested in situations that are theoretically appropriate. As time goes on, these instruments are so much taken for granted that they are incorporated in more complex pieces of apparatus as if they were perfectly simple components with purely practical and empirical functions.

Although it is not the normal practice of research to maintain a permanent distinction between observational and theoretical languages, this distinction might be possible in principle. Some philosophers have argued that this can be achieved by *reducing* all complex observational statements to simpler, empirical elements. The scientist is advised that metaphysics can be avoided by specifying the primary facts of science in terms of directly experienced events such as 'pointer readings'. Thus, the statement that 'the temperature was 12.75 degrees Celsius' means no more than that 'the pointer in the galvanometer connected to the thermocouple stood at the figure 12.75', or, even more primitively, 'a long black object stood in the white patch between an angular black mark and a curved black mark', etc., etc. This policy is, of course, quite impractical, because of the immense circumlocutions required to describe the simplest scientific event: but it might, in abstract principle, lead scientific knowledge back to a bastion of empirical facts, impregnable against the assaults of philosophical scepticism.

There are different versions of *positivism*, depending on the type of observational fact that is held to be primary. *Phenomenalism*, for example, would build scientific knowledge from reports of direct observations of 'phenomena', such as chemical reactions or collisions between particles, excluding interpretative concepts such as 'atoms', 'molecules' or 'forces'. *Operationalism* is a physicist's variant of this, emphasizing that a physical quantity is essentially defined by the operations needed to measure it. In psychology, the objective of *behaviourism* is to describe the observed behaviour of human beings and other animals without reference to internal mental processes.

Although positivist philosophies of science are opposed to 'metaphysics', they are not intended to be anti-theoretical. In any branch of science where unconvincing hypotheses have proliferated it is often very salutary to imagine that one could burrow down to a very elementary level of factual description, as a foundation for a more parsimonious theoretical scheme. For example, Einstein's General Theory of Relativity, with its emphasis on 'events' observed by different observers moving at different speeds, owes much to the phenomenalism of Ernst Mach. In practice, however, if the data are too fine-grained and prolific, it is impossible to see the regularities in them, and thus to go on to explain them. The working scientist is bound to group these elementary 'facts' into larger classes (§2.4), which soon become objectified and treated as empirical categories in their own right. The positivist stance is too rigid as a philosophy of research, because it has no place for the invention and validation of general laws and concepts, and thus takes no account of explanation as the fundamental goal of science.

This critique of positivism from the point of view of scientific practice does not necessarily invalidate it as a general philosophy. But such a philosophy cannot be constructed on some notion of a substratum of empirical facts that are 'scientifically' true. Careful analysis shows that even the most elementary 'phenomena', 'operations' or 'behavioural events' are just as theory-laden as the ordinary data of science. Even the most radical reduction of all human experience to brief episodes of the perception of primary *sense-data* – 'this, red, now' – is not defensible, because it embodies an obsolete theory of human perception. For example it fails to take account of the *tacit knowledge* of the circumstances, objects, events and operations surrounding and permeating all conscious human activity (including scientific work) which is impossible to catalogue and communicate in finite terms. It may be, indeed, that instead of trying to construct a general theory of our knowledge of the external world on a positivist model of scientific empiricism, one should turn to the philosophy of *phenomenology* for deeper insights into scientific epistemology. But this would take us too far from the philosophy of science as normally conceived.

3.4 The problem of induction

In its progress from description (§2.2) to explanation (§2.10), science moves from a realm dominated by facts to a realm of theory. But whereas facts are individual and particular, theories are presented as general and universal. This brings us to one of the main difficulties of justifying scientific knowledge.

The difficulty is that a general proposition such as '*all* swans are white' does not follow by rigorous logic from a finite number of particular instances of the form '*this* swan is white'. Since the general proposition cannot be tested empirically for all its possible cases – e.g. for swans not yet born – it could turn out to be false. Indeed that was exactly what happened when black swans were discovered in Australia. This is the famous problem, first posed by David Hume more than 200 years ago, and still an obstacle to the reduction of scientific knowledge to the same status as empirical reality.

It is clearly a genuine problem, since there are numerous cases in the history of science where even widely confirmed empirical generalizations have later been disconfirmed by contrary instances. For example, the law of the permanence of chemical atoms was later disconfirmed by the discovery of radioactivity. It also applies throughout science, far beyond the elementary case where an observed regularity or invariant pattern of facts is reformulated as a general 'law' (§2.9). All scientific theorizing is abstract and generalized, applicable in principle to whatever instances may actually occur. But a theory can only be validated empirically by an appeal to evidence that is finite and specific: there is always the possibility of turning up fresh evidence which the theory does not fit.

'Hume's problem' seems so direct and obvious that it is difficult to accept that there is no formal way round it. It now seems clear that the *induction* of general propositions from a finite number of particular instances is not of the same *logical* status as the inverse process, the *deduction* of particular instances from a general proposition. Thus, if, indeed, *all* swans *were* white, then it would be perfectly certain that *this* swan would be white – but the argument is unsound in the reverse direction. Although logical rigour is of inestimable value at many stages in the discovery and validation of scientific knowledge, it cannot be used as a rigid two-way connection between a superstructure of theory and a foundation of empirical fact. There have been many attempts to define further logical, extralogical, or metaphysical principles which would get over this difficulty, but these all turn out to have the same defect in some more abstract form. *Inductivism* is not satisfactory as a fundamental scientific epistemology. There is always this final element of insecurity and uncertainty in the justification of scientific knowledge.

3.5 Inference

Faced with the logical impossibility of justifying induction, science has had to develop intellectual strategies that do not pretend to arrive at absolutely certain conclusions by formal argument. Indeed, once it is admitted that all scientific knowledge is to some degree uncertain, there is much more room for epistemological manoeuvre. In particular, the extent of this uncertainty can be estimated.

The notion that scientific 'truth' need not be an 'all or none' attribute leads inevitably to the concept of *probability*. In everyday parlance, a general proposition that cannot be proved to be true is often said to have been shown from the evidence to be 'highly probable'. Although this is a confession of uncertainty or ignorance, it also suggests a high degree of confidence that the proposition will prove worthy of considerable further attention. For most practical purposes, it would be quite sufficient to show that scientific propositions are so very, very probable that they might as well be treated as completely true.

Can this familiar manner of speaking be explicated more rigorously? An elaborate calculus of probability has been developed for the statistical analysis of data in many fields of research; can this calculus be applied to the general epistemological problem of the *inference* of a general proposition from a finite set of particular instances? Can Hume's problem be dealt with by this device?

The basic philosophy of probability is notoriously murky and controversial, so that this programme attracts as many sceptical objections as it is intended to allay. Nevertheless, a probabilistic attitude to scientific epistemology is so widely held that it is instructive to work through a few elementary cases to uncover its strength and weaknesses. Suppose, for example, that we had examined 100 swans' eggs that had broken before hatching, and found that they all contained excess quantities of DDT. It would be natural to infer that 'DDT always causes swans' eggs to break before hatching'. To what degree is such an inference justifiable?

This would appear to be an ideal case for the application of the *Law of Succession*: 'If all that is known of an event is that so far there have been n occasions on which it could have occurred, and that it has happened on all of them, then the chance of it occurring on the next possible occasion will be $(n+1)/(n+2)$'. According to this theorem, the probability that the next broken swan's egg will be found to contain excess DDT is thus $101/102$ – i.e. greater than 99%.

In many ways, this is a reassuring calculation, showing that formal statistical analysis is reasonably consistent with practical rationality. In most fields of science, any generalized proposition having this degree of certainty would be regarded as 'highly significant' so that we are justified in our intuitive judgement that the evidence for it is 'very convincing'. Nevertheless, the number obtained by this calculation is not to be trusted as a measure of any well-defined quantity.

In the first place, it does not point uniquely to the inferred generalization. Our observations would obviously be consistent with many other, more complicated propositions, such as a *statistical* association between excess DDT and egg breakages. A simple 'law' that DDT *invariably* has this effect is quite different in principle from the proposition that this phenomenon occurs, on the average, in only 99 cases out of 100. It would take a much more elaborate investigation, on a much larger sample, to narrow down all such possibilities until they could be effectively disregarded. Indeed, the main practical use of statistical analysis is not to justify *likely* hypotheses, but to eliminate the very *unlikely* ones. In this case, for example, one such hypothesis might be that excess DDT has *no* effect on the viability of swans' eggs, so that the observed association is just accidental: it is easy to prove, under certain plausible assumptions, that the probability of finding 100 successive cases like this in a random sample is in the order of 1 in 10 billion, and can therefore be discounted. In practice, this sort of argument is applied to much more messy and 'noisy' data, where it often requires very elaborate mathematical analysis to decide whether an observed *correlation*, or other statistical regularity, could have arisen more or less by chance, without some conjectured causal connection.

But the prime objection to using the Law of Succession, or any other formal probability calculus, to justify an inductive inference is that it is almost impossible to allow for all the tacit knowledge concerning the situation. To what extent can it be said, for example, that 'all that is known' of an event is that it has occurred or could have occurred. The investigator studying the effects of DDT on swans' eggs would surely have taken various elementary precautions to avoid systematic bias in his observations. He or she would have checked that unbroken swans' eggs rarely contain 'excess' DDT, or that the sample was not drawn from a region where the whole environment was polluted with this substance. But all such precautions imply considerable *a priori* knowledge of the nature of the events being studied, thus contradicting the terms of the theorem. In principle, the characteristics of the 'events' to be observed could be redefined to allow for some of this knowledge, but this process is endless. The reformulated calculation quickly becomes so elaborate, with so many extra imponderable factors to allow for what is already known, that the numerical result is meaningless.

The same difficulties arise, even more disconcertingly, when attempts are made to quantify the contributions of *a priori* tacit knowledge to the more theoretical terms in the inferred proposition, such as the likely existence of a 'causal' connection underlying the observed association of events. It might be suggested, for example, that excess DDT is produced as a byproduct when swans eat some other food that has this effect on their eggs, so that the connection is not causal at all. The plausibility of the inferred generalization would then depend on how much was already known of the metabolism of this compound and much other biochemical and physiological knowledge. Although the generalization might then be reformulated as a mere statistical correlation of observed 'facts', this positivist reduction must eventually fail for the reasons already discussed in §3.3.

From a formal point of view the fundamental objection to a strict probabilistic account of induction is that the process of inference does not itself satisfy the fundamental axioms of the mathematical theory it attempts to apply. The probability calculus refers primarily to the relative frequency of occurrence of various discrete items in numerous runs or re-runs of the same basic processes, such as throwing dice. These circumstances are often extended imaginatively to very complicated situations, but they do not really apply to the act of asserting a general proposition on the basis of finite evidence. The 'items' that would have to be counted are themselves alternative hypotheses – that is 'ideas' – whose discreteness or equivalence cannot be measured, whilst the process as a whole cannot be re-run several times to see how often the same overall result is obtained. Thus, the number coming out of any such calculation is purely notional, and can only vaguely represent the 'extent of the uncertainty' in the knowledge in question.

This theme eventually returns to its starting point, which is the familiar usage of 'probability' to represent some relative degree of 'truth', or of 'rational belief'. This is not an empty usage, for it does approximately mirror the stance that is normally taken to events where chance really does play a significant role, but it cannot be axiomatized to yield objective quantitative results. Even when it is meaningful to say that, on the given evidence, a particular hypothesis is very *im*probable, the odds against it can only be estimated very roughly. The probabilistic approach to scientific inference is instructive as a metaphorical scheme of justification that is more or less consistent with research practice, but it does not provide a rigorous, epistemological 'method' for science as a whole.

3.6 Prediction

The most convincing demonstration of the validity of a scientific hypothesis is the successful *prediction* of a previously unobserved or unrecorded phenomenon. Historical examples, such as the discovery of new chemical elements with the properties predicted by Mendeleev on the basis of gaps in his Periodic Table, are so striking and persuasive that it sometimes seems as if this is the only way in which scientific knowledge can be properly validated.

It is a serious oversimplification however to equate science with 'predictive rationality'. It is true that scientific knowledge tends to exclude discussion of events that are so irregular and 'unpredictable' that they cannot be classified by any known principle. The *practical* importance of scientific predictability is also obvious, for this is ultimately the means by which science gives people *control* over nature and over other people. To undertake any rational project, such as designing a machine, destroying an enemy, or curing a disease, it is essential to have some idea of the consequences of one's actions. Most scientific research is, in fact, consciously directed towards just such objectives (§12.1).

But these pragmatic considerations do not fully cover the intellectual goal of science – that is, *understanding*. Prediction is not the necessary and sufficient condition for scientific *explanation* (§2.9). There are, for example, perfectly sound 'historical' sciences such as geology, palaeontology and cosmology, whose phenomena cannot strictly be predicted since they all occurred in the past – although theories about these phenomena may be satisfactorily validated by the discovery of 'predicted' evidence after they were originally proposed. The relationship between prediction and explanation is not symmetrical or transitive. Thus, for example, we understand how earthquakes occur, and can explain the mechanism of hurricanes, without being able to predict them. Conversely, the successive stages of an incurable disease can often be 'prognosticated' – i.e. predicted phenomenologically on the basis of experience – without a satisfactory explanation of how and why this happens. Genuine predictive power is too distant an ideal, and too narrow an indicator of progress, for science in general.

Epistemologically speaking, however, the success of a scientific prediction must surely strengthen the credibility of the hypothesis on which it was based. Exactly why this should be so calls for very deep analysis, since the notion of scientific credibility can scarcely be defined without some broad reference to the outcome of consciously directed action. By the same tokens, it is difficult to think of any way of deliberately validating a scientific theory that does not involve the empirical investigation of future circumstances implied by that theory. Thus, the outcome of an experiment can only be judged in relation to what was predicted before it was undertaken. But this does not mean that a hypothesis is made certain when its predictions are confirmed.

This follows at once from the impossibility of proving the truth of an inductive generalization (§3.4). Suppose, for example, that a general law has been inferred on the basis of 100 positive cases, with no exceptions. For this law we may confidently 'predict' that the next case will also prove positive, and try the appropriate experiment. But if it succeeds, all that we have is a sample of 101 such cases, which only slightly increases the credibility of the law in question. According to the Law of Succession (§3.5), for example, the 'probability' of our inference has increased from 101/102 to 102/103, which is theoretically and practically quite inconsequential.

Indeed, this result could be misinterpreted as supporting some underlying theoretical mechanism which we are trying to test, when in fact it is no more than a little further evidence for a phenomenological regularity that is already adequately established.

On the other hand, it must be admitted that if this particular test were to *fail* – if the predicted result were *not* obtained – then this would be extremely significant. The whole theory on which the prediction was made could be put in jeopardy. Indeed, in strict logical terms, a single *disconfirming* instance – e.g. a single observation of a black swan – can empirically *falsify* a previously undoubted generalization – i.e. that *all* swans are white. Scientific propositions are never so well-posed that this sort of logic can be rigorously applied, but the general effect is clear enough. A probabilistic calculation would show, for example, that the 'probability' of the generalization would have dropped from near unity to near zero, as a result of this single observation.

This is the epistemological point that was seized on by Karl Popper to show that empirical falsifiability must be an intrinsic property of all scientific theories. This principle goes beyond asserting that every inductive generalization is logically fallible (§3.4) and that scientific hypotheses can be *corroborated*, but can never be absolutely confirmed by successful predictions; it emphasizes that any tests aimed at validating a theory (§2.14) must be genuine in that they must have the distinct possibility of giving a disconfirming result.

Popper's principle is extremely influential in the contemporary philosophy of science (see §3.7). It also suggests a rationale for the research strategy of testing theoretical predictions that otherwise seem to be *unlikely* to be confirmed – for if, against the odds, the trial is successful, the credibility of the theory will be changed from a small value to something approaching certainty. This is illustrated by the story of Mendeleev: when he proposed his Periodic Table, it was quite likely that several new chemical elements would in due course be discovered, but the probability seemed very small indeed that out of all possible physical and chemical characteristics they would have just the properties he predicted. The corroborative effect of such episodes can easily be explicated by a formal application of probability theory, relying upon an axiomatic principle known as *Bayes' Theorem*. This demonstrates that the '*a priori* probability of a hypothesis' can be increased in this way by a large factor, but gets no nearer to giving an objective meaning to this elusive concept. Indeed, by drawing attention to the immeasurability of the alternative futures that have to be taken into consideration in such a calculation, it shows clearly that no formal analysis of inductive inference can be more than a metaphorical schematization of human intuition, even in the highly rational practice of scientific research.

3.7 The hypothetico–deductive method

In reality, most scientific theories are much more complicated than 'laws' (§2.9). Various hypotheses are built into or around elaborate 'models' (§2.11), closely interwoven with verbalized relationships, diagrams and mathematical formulae. Nevertheless, the process of justification can often be separated into two phases, corresponding to the elementary notions of 'prediction' and 'confirmation'.

The first phase is essentially 'theoretical'. In principle, a theory should be a rationally articulated structure of clearly defined concepts (§2.12) and ought therefore to be open to formal logical analysis. Verbal and/or mathematical argument can therefore be applied to extend its scope, and to *deduce* various empirically observable consequences. Scientists often say that the theory is used to 'predict' certain facts, even when some of these are already known, and may even have been what the theory was actually created to explain.

This is where *logical rationality* is paramount in science. Any ambiguity, imprecision or inconsistency in the chain of argument leading from the hypothesis to its implications can give rise to serious errors or uncertainties in the validation process. Thus, for example, acceptance of the Theory of Continental Drift was delayed for nearly half a century by failure to appreciate that certain calculations about the rigidity of the Earth were crucially dependent upon quite unverified assumptions about the behaviour of rocks at enormous pressures and temperatures. In this phase, the relative merits of rival hypotheses as sources of testable implications are an important consideration. In general, the hypothesis that is most *informative*, in the sense of making the most definite yet unlikely predictions over the widest range of ascertainable facts, has the best chance of being either triumphantly corroborated or decisively disconfirmed (§3.6).

In the second phase, the question is: 'Do the facts fit the theory?'. This is, of course, the realm of *experiment* (§2.8), since it is usually necessary to undertake a special investigation to bring to light some of the unknown facts predicted by the theory. This phase is dominated by *practical rationality*, epitomized in the use of sensitive *instruments* (§2.6) to make accurate *measurements*. Thus, for example, Einstein predicted, from his General Theory of Relativity, that the path of a ray of light passing very close to the Sun would be very slightly 'bent'. This tiny effect could only be confirmed by very precise astronomical observations of the apparent positions of the stars, at the moment of a total eclipse of the Sun. Indeed, this was a case of a *crucial* experiment, since it seemed to discriminate clearly between Einstein's theory and other theories of gravitation, such as Newton's, which would not have predicted this particular phenomenon. The design of an experiment to test a theory is one of the major arts of science, since it calls for a clear understanding of the theoretical conditions under which a novel phenomenon has been predicted,

3.7 The hypothetico-deductive method

and ingenuity to manipulate nature to set up these conditions and observe the outcome.

This two-phase process of validating hypotheses is the essence of the *hypothetico-deductive method*, which has long been the standard philosophical model for the 'scientific method'. It is obvious that this is merely an extensive elaboration of the elementary policy of setting up a generalization by induction from a regular pattern of empirical observations, and then seeking confirming instances of its predictions – in a word, of the policy of 'trial and error'. From the point of view of an epistemological purist, it is no stronger than that policy (§3.6): when the facts do, indeed, fit the theory, they can give a powerful boost to its credibility, but they can never validate it absolutely: when a prediction fails, the theory is 'falsified', and ought perhaps to be discarded. In the face of Hume's problem, and various other philosophical difficulties, there is not much more that could be said.

Considered more broadly, however, this account of the 'method' of science has the great merit of suggesting a *demarcation criterion* between scientific knowledge and other forms of organized thought. Popper's provocative formulation of this criterion as 'potential falsifiability' emphasizes that scientific theories must be capable of being *genuinely* tested against empirical reality, which is at the very heart of the hypothetico-deductive method. In other words, science seeks out situations where there is direct contradiction between theory and experience, and then modifies theory in the light of experience: it has no place for certain types of grand theory which are so vaguely formulated and held to be so true in essence that it is experience that must be 'reinterpreted' to fit them. This is a most valuable precept, even though a thorough analysis of scientific thought and practice might well reveal that science itself is founded on 'metaphysical' principles that do not satisfy this criterion.

On the other hand, the actual method of science is not adequately characterized as a chain of 'conjectures and refutations'. This formulation does not allow sufficiently for the high credibility of well-corroborated theories, nor for the provisional acceptance of partially falsified hypotheses. The point is that neither 'facts' nor 'theories' are as logically clear and distinct as is assumed in the elementary accounts of 'prediction' and 'confirmation'. They interpenetrate each other's domains (§3.2), and are hedged around with tacit circumstantial and existential assumptions that cannot be rigorously taken into account when one is set to 'test' the other. In practice, when a prediction has failed, it is almost always possible to think of a good reason why this should have happened, without abandoning the hypothesis one is supposedly testing. It is part of the art of scientific investigation not to be daunted by a few negative experimental results, and not to reject hypotheses that seem to have been 'falsified' by a few disconfirmatory instances but which still have a great deal going for them in other ways. But this sort of objection to the hypothetico-deductive model of science would apply to any attempt to idealize the *practical* epistemology of research, where the possible errors and uncertainties are far more evident than would be reported in retrospect to the philosophical analyst.

Perhaps the greatest merit of the hypothetico-deductive model is that it provides a rationale for the *dynamics* of science as a whole (see chapter 7). In the context of discovery, this may be conceived as a progression from the empirical to the theoretical: in the context of justification, the cycle closes by a return to 'facts' to confirm the theories. This explains the continued interaction of theorizing and experimenting. Experiments are given purpose by the theories they are designed to test; new theories are needed to explain the results of experiment. In the course of this interaction, 'problems' define themselves (§2.14) and become the focus of intensive investigation until they are solved. Theoretical understanding grows, and is continually refined by the factual refutation of conjectures, which provides in turn an incentive for the generation of new hypotheses.

But here again, this philosophical account of scientific activity should not be taken as definitive. The hypothetico-deductive model only describes research in the context of justification: it says nothing about how hypotheses should be brought to mind in the first place, and is embarrassingly silent on the very common phenomenon of serendipitous discovery (§2.5). In other words, it ignores the contributions of *imagination* and *curiosity* to the growth of knowledge. It is also inadequate as a basis for research policy in certain areas of science, where *exploratory* investigations, without strong theoretical presuppositions, may be called for. Thus, it would be quite inappropriate to apply a rigid hypothetico-deductive programme to a science such as hominid palaeontology, where fragmentary information is accumulated more or less at random, and has then to be interpreted as evidence for or against various equally fragmentary theoretical schemes. Such disciplines are as much part of the scientific enterprise as are theoretical physics and molecular biology.

3.8 Established knowledge

Philosophers and scientists alike tend to be much more interested in the research 'frontier', where discoveries are being made and tested, than in the more settled regions of knowledge whose validity is no longer seriously contested. But science purports to tell us what is *known*, not what is merely conjectured. The goal of research is to create a body of knowledge that is so well *established* that it can claim to be beyond serious doubt. Although this claim can always be challenged, it is interesting to consider the characteristics of scientific knowledge in its mature state.

Whether or not it is really true (§3.9) a mature science must be, above all, *reliable*. If a well-founded body of knowledge predicts unusual phenomena, these should be observed: if it is used to solve a technical problem, that solution should work. The pursuit of knowledge does not always have practical goals, and pragmatism is not

3.8 Established knowledge

the touchstone of scientific credibility, but a science that turns out to be unreliable in use will eventually have to be rejected or radically reformed.

The fact that science 'works' is as important for the scientist in the laboratory as for society at large. The whole body of scientific knowledge is so enormous that it can only be properly understood very locally by the individual researcher (§5.3). Every scientist must therefore rely completely on the validity of all the pieces of 'established' knowledge that he or she must use incidentally in the course of research. Any experiment in physics, for example, would be entirely frustrated if the researcher had to stop and test every basic theoretical principle and empirical result that had gone into the design of the apparatus, from Maxwell's equations for the electromagnetic field to the calibration chart of the thermocouple amplifier. Every research project not only stands on a foundation of established knowledge in its own domain; it is also deeply embedded in bodies of established knowledge drawn from other disciplines whose validity the researcher is quite unable to assess.

Outside of their own narrow specialties, therefore, scientists tend to treat established knowledge as an indubitable description of nature, to be taken on trust with whatever else they believe about the everyday life-world. This description is not, of course, a complete picture of reality, including every contingent datum or fact: it is a *schematic* representation, constructed symbolically or metaphorically out of general concepts and their theoretical relationships. A mature body of scientific knowledge is something like a *map*. The structure of some region is represented by the relative positions of various conventional symbols, each standing for some selected category or aspect of the real world. The information it can give is not, so to speak, concrete and immediately comprehensible: it has to be manipulated mentally and interpreted before it can yield particular facts. But like a good map, a well-formulated scientific theory can be an inexhaustible source of empirical facts which could never be separately listed or otherwise grasped as a whole.

The map metaphor also suggests that scientific knowledge is a *multiply connected network* of concepts, where the validity of any particular proposition does not depend solely on one or two other theoretical propositions or empirical observations. The history of science tends to focus on the discovery and primary validation phases in the life of a theory, when it was still hypothetical and could easily have been rejected as a result of a crucial experiment or a cogent critical comment. But as soon as it is widely accepted, it accumulates numerous links with other theories, so that it can no longer be amended without repercussions for many other items of knowledge, theoretical or empirical. Thus, for example, anyone who would now set out to refute relativity theory could not succeed by, say, demonstrating a logical fallacy in Einstein's original paper. It would also be necessary to explain the successful use of this theory in a vast range of scientific work, ranging from the engineering design of particle accelerators to the fundamental equations of quantum physics. Any change in the essential results of such a well-established theory would have incalculable consequences far beyond the facts it was conceived to explain (cf. §16.3).

The close-knit theoretical scheme of a mature science must therefore be logically *self-consistent*. If the connections between its various conceptual components have not been properly made, this would soon show up as a logical contradiction or ambiguity, making the whole scheme unreliable in practice. Since theories have often germinated in different phenomenological domains, and grown outward until they made contact – witness the conjunction of physics, chemistry and biology in the field of molecular biology – the search for possible internal inconsistencies is a significant research programme in many mature scientific disciplines.

In a discipline such as physics, where theory is built around sophisticated mathematical models, it is essential to *axiomatize* these as precisely and as economically as possible, in order to check their formal consistency. This programme of precisely defining every basic assumption of the theory, in the traditional spirit of Euclidean geometry, offers scope for advanced mathematical analysis, with the same attention to rigorous proofs as in pure mathematics. But a highly axiomatized formulation is not necessarily the ideal way of 'mapping' an aspect of nature: what is gained in definiteness and outward logical consistency must be balanced against loss of significant empirical detail in the process of formal abstraction. In almost all fields of biology, for example, a good sketch diagram may be no less valid epistemologically, and much more comprehensible, than a whole book of equations. There is no general philosophical demonstration that the final, most mature state of all science must be similar to theoretical physics in this respect.

But the characteristic multiple connectivity and self-consistency of any established body of scientific knowledge can be very confusing at the research frontier. The question then becomes: what can be taken as given, and what is still conjectural? Even within a tightly woven theoretical scheme, where the research goal may be no more than a modest extension of the scope of the theory, or a reformulation of the relationships between the concepts, it becomes difficult to decide which propositions should be regarded as the *premises* of the argument or calculation, and which are the *postulates*. Where the empirical validity of the knowledge is in question, there is even more confusion about which items should be treated as 'theoretical' – hence in principle corrigible – and what are indisputable 'observable' facts (§3.3). For a scientist looking for very rare phenomena such as solar neutrinos, or quarks, or gravity waves, this uncertainty can spread back into the realms of supposedly well-established knowledge, and become the central issue in the interpretation of every investigation.

Away from the research frontier, these doubts soon vanish: all well-established theory is taken to be valid, and all that can be deduced from it is treated as an empirical 'fact'. But even though there is no incentive to observe such a fact directly

3.8 Established knowledge

so as to test the theory further, its epistemological status does not stay constant. On the one hand, every theory that contributes in part to any prediction is corroborated by the success of that prediction. The atomic hypothesis of chemistry, for example, has been tested in principle and confirmed in practice by every successful experiment that was designed on its basis, ever since the early nineteenth century. Conceptual principles are continually being corroborated by their contribution to the design of successful technological devices: could one imagine a more convincing test of Newton's laws of motion and Law of Universal Gravitation than sending a space vehicle in free flight from planet to planet? Thus, certain pieces of scientific knowledge can become more and more reliable, more and more credible, until they really are *established* beyond all possible doubt.

On the other hand, *all* scientific knowledge is in principle *corrigible*, as was demonstrated, on an immense scale, by the overthrow of the firmly established geological principle that the continents were fixed in their positions on the globe. Such spectacular scientific revolutions (§7.3) are fortunately rare; but scientific knowledge is continually being revised in detail.

Scientists are seldom as scrupulous as they ought to be in testing all their hypotheses beyond reasonable doubt, and even make quite radical assumptions that they fail to test at all. As every research worker knows about his or her *own* field, there are many weak arguments to strengthen, loose ends to tie up, and even 'floating' pieces of accepted theory that are not really connected into the network at all. To recall a striking example: it took many years for physicists to realize that a very important general principle of quantum theory – the 'Law of Conservation of Parity' – had been taken for granted, without any independent test; experiment showed that this so-called 'Law' was not true at all. Thus, what looks like a coherent well-established body of knowledge is seldom as closely connected and tied down empirically as the outsider is led to believe.

Whether by practical application, by radical revolution, by minor correction, by the reformulation of conceptual schemes, or by the accumulation of new results round the edges, scientific theories are continually changing in their content and credibility. The hypothetico-deductive method itself is like the process by which biological variants and mutants are 'selected' for their fitness. It thus implies an *evolutionary* epistemology of science as a whole (§7.4). Out of a multitude of conjectures, only a small proportion survive the ruthless process of empirical validation. When we speak of *established* scientific knowledge, we should not be thinking of it as an eternal verity, permanently fixed in the precise form in which it is currently taught, and believed, and put to use. We cannot even be sure that it is converging on some unique state of 'ultimate truth', to be revealed to mankind in a far-distant future. Although epistemologists have been trying for centuries to prove that science ought to be perfectible in that respect, they have not, so far, succeeded.

3.9 Does science describe reality?

The scientific description of the natural world is very impressive. For those aspects of life that it claims to cover, it is as reliable as any knowledge that we have, at present, to guide us. We are all philosophers, however, in asking whether this description is 'true'. Does science describe the world 'as it really is'?

This question cannot be answered properly without a deep plunge into *ontology* – the branch of philosophy concerned with the meaning of 'existence' in the broadest sense, including, of course, the existence of everyday life-world objects such as screwdrivers, as well as more paradoxical entities such as unicorns or charmed quarks. But from what has been said so far, it is obvious that the answers given by most philosophers must lie somewhere along a line extending from extreme *realism*, which emphasizes the *factual* content of science, to the opposite pole of *conventionalism*, which stresses the *theoretical* characteristics of scientific knowledge.

The basic argument for *realism* is that science is grounded firmly on empirical observation (§3.2). The research process enables us to discover distinct, coherent entities, such as atoms, or viruses, which later turn out to be just as real as tables or chairs, even if they cannot be directly seen or felt. Our notions about such entities are at first very misty and vague, but they acquire substance and clarity as their existence is corroborated by further investigation. Of course we might be mistaken, in our initial scientific conjectures, just as we often are in everyday life, but that is no reason why a well-established 'scientific' entity (§3.8) should be denied the ontological status that we give to any common-sense life-world object with which we happen to be familiar. From this point of view, scientific knowledge might even be considered more real than reality itself, because it has been so thoroughly tested.

This obviously extends the notion of reality far outside its normal limits. Some of these extensions are justifiable both scientifically and philosophically. For example, modern scientific thought is no longer prejudiced towards *materialism*; there is evidently more to existence than 'matter' and 'void'. To the physicist or engineer, the reality of the invisible quantity called 'energy' is far less questionable than the reality of the quantity called 'money' that is represented by a number in his or her bank account. Many theoretical entities seem to have little connection with the world that we know. The neutrino cannot be localized in space, probably has no rest mass, and interacts so weakly with 'real' matter that it could pass through a million miles of solid steel without being perceptibly scattered or captured: nevertheless, the evidence for the existence of this particle is as circumstantially sound as the evidence for the former existence of, say, Alexander the Great. What we call common-sense reality is not derived solely from what is immediately apparent to the eye and ear, but is also constructed by inference from all manner of other items of information.

Nevertheless, the irreducible corrigibility of scientific knowledge, the incessant change in its contents (chapter 7), and the coexistence of theories claiming to represent just the same facts, all seem to suggest that theoretical scientific entities lack the permanence and uniqueness we normally attribute to real things. At what moment, one might ask, did 'genes' or 'viruses' become real? At a certain period in the history of science, these were mere conjectures, of no more credibility than many competing concepts or models such as 'germ plasm' or 'noxious vapours'. Were the latter also 'real' at that time – or are we permitted the luxury of hindsight in pronouncing them to be fictions? With such historical examples in mind, how can we assert the reality of any contemporary theoretical entity that might equally turn out to have been a figment of a strong scientific imagination. In some branches of science, quite different theoretical schemes have to be used to describe different ranges of phenomena. Thus, in vacuum-tube electronics, the electrons are treated as particles, whilst in good crystalline solids they must be treated as waves. Although these schemes can be reconciled in principle in the formalism of quantum mechanics, it is hard to believe in such a contradictory 'reality'.

Although a systematic realist could counter such objections, they have driven many philosophers of science to the opposite ontological pole. *Conventionalism* assigns to scientific knowledge no higher status than that of being a useful hypothesis. The 'map' of science may be a most effective scheme for describing and ordering natural phenomena in an extraordinarily concise form (§3.8), but however well it is corroborated it remains essentially a fiction, whose entities are conventional symbols, not representations of real things in themselves. Indeed, this metascientific metaphor is particularly apt, for map-making is the science where the conventionality of natural knowledge is most clearly evident. No flat map of the surface of the round earth can show the shapes of large countries as they 'really' are: there are merely a number of conventional schemes, such as Mercator's projection, any one of which may be chosen for its practical convenience. At a more profound level, the geometry of space itself was once thought to be 'really' just as Euclid axiomatized it in 300 B.C.: in the early nineteenth century, mathematicians showed that it could be mapped according to all sorts of other schemes which are all equivalent in being consistent with the facts of physics.

Conventionalism has the great social virtue of modesty. It effectively debunks the doctrine of *scientism*, which asserts that science, and *only* science, can tell us what the world is really like (§16.4). This cautious attitude is even applicable from one science to another: on several occasions the physicists have dogmatically insisted that their 'more fundamental' mathematical analysis of the history of the Earth must be superior to the geological version deduced from the fossil record – and the physicists were wrong. Conventionalism also allows for continual change in the theoretical contents of science whether by new discoveries, by the correction of apparent errors, or by the consolidation and reformulation of existing concepts. It is thus consistent with all that history tells us about the growth of knowledge, both by peaceful evolution and by violent revolution (chapter 7).

The conventionalist point of view can be justified epistemologically by arguing that a scientific theory is never more than a metaphor (§2.11), explaining phenomena by analogy. A well-entrenched theory that has been successfully corroborated may have every desirable quality of simplicity, coherence and wide applicability (§3.8), and yet there is still no way of proving that it is more realistic than any other conceivable scheme. At the research frontier, it is common to find several alternative hypotheses that fit known facts about equally well, and even a hypothesis that seems to have been disconfirmed empirically need not be immediately discarded (§3.7) for it can always be modified arbitrarily to fit the new data.

The choice between competing models or analogies cannot be automated logically: it depends eventually on human judgement. In practice, it is often resolved by some 'convention', or tacit social understanding that only a particular scheme should be taught and used. Conventionalism thus leaves the door open for epistemological *relativism*. One of the basic principles of the *sociology of knowledge* (§8.2) is that different social groups quite naturally and properly adopt different conceptual conventions in their mappings of the facts of life.

Although conventionalism is fully defensible in principle, it obviously offends against the scientist's everyday feeling that the well-established knowledge permeating research is as real as anything could be. It fails to account for the vivid experience of research as *exploration*, where previously unknown realms of fact and concept are *discovered* – not merely *surveyed*, or artificially *constructed*. If this experience is illusory, so must be all our experience of the life-world. Scientific conventionalism has no defence against total philosophical *scepticism*. If scientific *theories* are arbitrary constructions, then so are scientific *facts*, into which theories are deeply interwoven (§3.3). But if *scientific* facts contain this arbitrary component, so do all *empirical* facts, including our most cherished notions of what is real. You question my 'scientific' assertion that I am observing an oviparous mammal, do you? In return, I can ask why you can be so sure that you are eating an apple. Total scepticism has, of course, a long and respectable intellectual history, but it cannot help us much in the philosophy of *science*, which must be concerned with the epistemological and ontological relationships between scientific knowledge and other forms of knowledge, especially of the everyday world.

The *positivist* programme tries to stabilize these relationships by defining scientific theory as a minimal set of generalizations about firmly established (i.e. 'real') sense-data, thus strictly confining conventionalism to the conceptual domain. Unfortunately (§3.3), this programme is always vulnerable to sceptical fundamentalism, which continually subverts the privileged status of what are supposed to be 'empirical facts'. On the other hand, if the nature and quality of these 'facts' is left out of the analysis, it is easy to fall victim to *rationalism*, which overemphasizes the part played by logic in science. Although theories must be logically consistent and rationally articulated (§3.8), they cannot be conceived intellectually without reference to the empirical tests by which they are most likely to be refuted. To avoid this criticism, the doctrinaire rationalist must take up the traditional philosophical position of *idealism*, which argues that the essence of *all* things lies in our ideas about them. Formal rationality plays an extremely important part in the construction and validation of all scientific knowledge, but it is a gross exaggeration of the logical rigour of scientific work to define its product as rationality incarnate.

3.10 Regulative principles of scientific work

There is no agreed philosophy of science, just as there is no agreed philosophy of life in general. This survey of the *philosophies* of science is perfunctory and inconclusive, as might be expected of any account of a thriving academic discipline. On almost every issue, the objections to each particular point of view seem to outweigh the supposed gains. Many of the controversial topics which have been hinted at here are not even complex or obscure in principle: they are just extraordinarily difficult to resolve. The excellent introductory books suggested for further reading explain these issues very clearly, and also discuss a number of topics, such as *determinism, reductionism, holism*, and evolutionary *emergence*, which have arisen in general philosophy under the influence of scientific ideas.

But it would be wrong to conclude from this survey that the scientific enterprise lacks a 'philosophy' to justify its practices, even though it cannot absolutely guarantee the quality of what those practices produce. This 'philosophy' is, of course, *metaphysical*, in that it cannot be derived from other undoubted propositions based upon experience. It is *regulative* rather than *prescriptive*: scientific work may be *characterized* by certain general principles, but is not *limited* or *defined* by them, as if they were rules or laws.

The first principle asserts the existence of an *external world* which is the primary concern of science. It thus dismisses extreme versions of *solipsism* and *subjectivism*, which lay their emphasis upon the internal thoughts and emotions of the individual person. But this principle does not endorse simple realism, since it does not assert that scientific knowledge is a true, or even unique, account of this world.

It is essential for scientific work that there should be some discernible *regularity*, or *order* in what is observed. Science would be impossible in an entirely chaotic universe, where objects were as impermanent, and events as random, as they appear in dreams. Or, to assert this principle in its weakest form, the scientific approach applies only to those aspects of reality that can be patterned according to general laws (§2.4) which do not vary arbitrarily from place to place and from time to time.

The third principle might be that the external world explored by science is *not disjoint*. Although scientific theories of various types evolve independently from observation and experiment on apparently unconnected aspects of human experience, they often extend towards one another until, in relation to some facts, they appear to overlap. Under such circumstances, they must be able to connect with one another without logical or empirical contradiction – or else they must be capable of revision until they can be united conceptually. This principle is essential if the search for improved scientific *explanations* (§2.9) is not to be forbidden by the very nature of things, but it does not say that the process of unification is easy, or even practicable in all cases. Thus, it is not a warrant for a strong *reductionist* programme, by which all scientific knowledge could supposedly be 'reduced' to certain fundamental theoretical concepts.

If science is to be considered a *progressive* enterprise (§7.1), there must also be faith that the effort to describe some aspects of the external world by a scientific 'map' (§3.8) is not a vain or ephemeral endeavour. Research is undertaken for a variety of motives – practical, aesthetic, or moral – which need to be gratified by the personal impression that what it produces is 'useful', or 'beautiful', or in some other way valuable. In other words, science is not simply a recreation, where the activity is its own reward, but is believed to create entities of permanent human significance (§16.6).

But epistemic precepts, such as the 'Principle of the Uniformity of Nature' that was traditionally used to justify scientific induction (§3.4), are inadequate to characterize an activity which is *collective*. As the last of the above principles begins to suggest, an *individual* 'philosophy' of science cannot be expressed without reference to the *social* dimension of scientific work and its products. The philosopher's discussion of epistemological and ontological questions needs to be extended and complemented along the lines already sketched out in §1.5. We need to study the *social relationships* of the people who are involved in research, drawing attention to the role of *communication* (chapter 4) in the construction and criticism of theories, and of intersubjective *consensus* in the consolidation of a body of empirical 'facts' (§8.5). We may then see science as primarily a *social institution*, whose output belongs essentially to what Karl Popper has called 'world 3', a world of *public* knowledge (§8.6).

Further reading for chapter 3

An excellent elementary text is

    R. Harré, *The Philosophies of Science*. London: Oxford University Press, 1972

For a broad-based, undogmatic survey, read

    P. Caws, *The Philosophy of Science*. Princeton, N.J.: van Nostrand, 1965

Readers with a knowledge of physics would appreciate
> N. Campbell, *What is Science?* London: Methuen, 1921 (reprinted)

and
> J. Powers, *Philosophy and the New Physics*. London: Methuen, 1982

An elementary introduction to the application of probability theory to the problem of induction can be found in
> R. A. R. Tricker, *The Assessment of Scientific Speculation*. London: Mills and Boon, 1965 (pp. 20–63)

The full case for a probabilistic epistemology of science is given by
> M. Hesse, *The Structure of Scientific Inference*. London: Macmillan, 1974 (pp. 89–102, and in detail thereafter)

The tacit component of scientific knowledge is discussed at length by
> M. Polanyi, *Personal Knowledge*. London: Routledge & Kegan Paul, 1968

The problem of induction is dealt with in several works by K. R. Popper, especially
> K. R. Popper, *Conjectures and Refutations*. London: Routledge & Kegan Paul, 1963 (pp. 33–65)

A standard reference for anti-formalist philosophies of science is
> P. Feyerabend, *Against Method*. (1975) Reprinted London: Verso, 1978

but a much better-argued book along the same lines is
> A. Naess, *The Pluralist and Possibilistic Aspect of the Scientific Enterprise*. London: Allen & Unwin, 1972

which also contains a critique of Popper's theory.

A relevant case study is
> H. M. Collins, 'The replication of experiments in physics', in *Science in Context*, ed. B. Barnes & D. Edge, pp. 44–64. Milton Keynes: Open University Press, 1982

4

Communication

'However certain the facts of any science may be, and however just the ideas we may have formed of these facts, we can only convey false impressions to others, while we want words by which these may be properly expressed'. *Antoine Lavoisier*

4.1 The archival literature of science

The basic principle of academic science is that the results of research must be made *public* (§1.5). Whatever scientists think or say individually, their discoveries cannot be regarded as belonging to scientific knowledge until they have been reported to the world and put on permanent record. The fundamental social institution of science is thus its system of *communication*.

How can one get to know what is known to science? In its most primitive form, scientific knowledge is to be found in the *primary literature* of science. This is a vast collection of 'articles', 'papers', 'research reports' and similar documents usually in a very conventional style and format that dates back to the origins of modern science in the late seventeenth century. A *primary scientific communication* is an original contribution to knowledge, by a named author or authors, normally published as a *paper* or *article*, of limited length (up to 50 pages, say) in a *periodical*, or *journal* devoted to a specific scientific subject. In the past, original research results were often reported for the first time in *books* – for example, Charles Darwin communicated his theory of evolution in a long book, *The Origin of Species* – but in the natural sciences this is now much less the custom than it is in the social sciences and humanities.

These documents constitute an *archive*, from which particular items of scientific *information* – the result of an experiment or observation, the definition of a theoretical concept, tables of numerical data, mathematical formulae, photographs, maps, etc., etc. – must be *retrieved* for further research or for practical application. This is not a trivial matter. The primary literature of a major scientific discipline such as physics may amount to 100 000 individual papers each year, published in many different

4.1 The archival literature of science

countries in many different languages. A whole battery of *retrieval* systems has been devised to deal with this practical problem. From the beginning of the nineteenth century there has grown up a large *secondary* literature, consisting of *bibliographies*, '*abstracts*' *journals, data compilations, review articles*, and other regular publications, cataloguing or surveying the contents of the primary literature. In the past 20 years, these secondary services have been largely computerized: printed indexes are now usually compiled automatically, or even superseded by direct access from a user terminal to a stored data base.

The information retrieval system of science does not, of course, make any direct contribution to original scientific knowledge, but it does have a powerful influence on the form in which primary research results are communicated. For obvious practical reasons, primary papers must have standardized bibliographic characteristics. The title of the journal, for example, must be correctly recorded, or abbreviated according to an international code. The volume and page numbers are essential address tags, with the year of publication as a general indicator of priority of publication. For reasons that will become clearer in later chapters, the name or names of the author or authors always remain attached to the bibliographic reference, together with some information about where they live and work. And of course the title of a long paper tells one very little about its contents, which are therefore usually summarized in a brief 'abstract' published with the main text.

Armed with these bibliographic details, one can go into a scientific library and pull down from the shelves the bound volume containing the reference one wants to read. The real difficulty in coping with the primary scientific literature is to know where to look for the papers that are likely to contain a specific item of information. Suppose that we want to know the melting point of a particular chemical compound or the observations that might throw light on some particular theory of animal behaviour. The obvious first step is to consult a comprehensive *data compilation* or an analytical *subject index* covering the field in question. But scientific knowledge is not just a collection of distinct items of 'information' that can easily be arranged in well-defined categories: it is a network of facts and concepts (§3.8) connected in all sorts of ways by theoretical implications, empirical cross-references, observational correlations and shared experimental techniques. Thus, the melting point of the compound may only have been reported incidentally, in a paper devoted to a class of chemical reactions of which it happens to be a catalyst, and the relevant evidence on animal behaviour may be scattered through a vast collection of miscellaneous reports on all sorts of other subjects. Computer methods can now be used to pick up key words, or significant data, from the titles, abstracts or even from the full texts of primary papers, but there is no practical way of constructing a complete 'taxonomy' (§1.5) of what is 'already known to science'. The fundamental problem of information retrieval is still unsolved.

4.2 Linkage by citation

Scientific papers are not addressed to the general reader: they are written for, and read by, other scientists. The primary scientific literature is the formal linking mechanism between the research work of individual members of the scientific community. This linkage is in tension between cooperative and competitive forces.

Science is cumulative and *progressive* (§3.10). It is built very largely upon previous science, whether by extension or by critical reassessment. Hence every new contribution must make full reference to the facts and theories on which it claims to be based. Scientific papers are linked to previously published papers, which are formally *cited* as authoritative sources of these facts and theories. This linkage extends from author to author. For example, my citation, in 1969, of the well-known paper by P. W. Anderson (1958; *Physical Review*, **109**, 1492–1505) not only legitimated my use of his equations in my own research; it also indicated that we were, in effect, scientific colleagues, working on the same problem.

On the other hand, scientific work must contain, above all, an element of *originality*. The goal of every research worker is to make *discoveries* (chapter 2). Questions about what discovery was made, by whom, on what date are often hotly disputed, for these are the basis upon which the material and symbolic rewards of successful research are distributed within the scientific community (§5.1). By convention, the claim for *priority* in having made a particular discovery is assigned to the author or authors of the first published paper reporting that discovery, and runs from the date on which the editor of a reputable scientific journal receives the typescript. Thus, Phil Anderson's paper in the *Physical Review* in 1958 reported a theoretical discovery for which in 1977 he was awarded a Nobel Prize. By citing his paper in 1969, I was endorsing his 'property right' in this discovery, and perhaps giving him some small satisfaction that it was being put to use.

A modern scientific paper will often cite several dozen other papers, spread over a period of 10 years or more into the past. This is consistent with our philosophical understanding of scientific knowledge as a closely interconnected network of facts and concepts (§3.8) and our sociological observation that scientists belong to a strongly interactive community (chapters 5 and 6). The formal communication system of science makes these interconnections and interactions quite evident. By taking part in this system, scientists become aware that their own particular investigations and discoveries are deeply embedded in, and dependent on, the work of many other people.

At the same time, they acquire the critical stance required by the 'method' of science (§3.7). It turns out that not all 'discoveries' are valid, and not all citations can be favourable. Scientists soon learn that they stand in a tense relationship with all other scientists publishing papers on similar topics. Within this relationship they may hope that their contributions will win recognition by appreciative citation; they must also be prepared to see their work treated with no more respect than they show to the work of others – that is, they may see it disconfirmed, demolished theoretically, superseded, or (saddest of all) simply ignored (§4.6).

Citations are published in such a precise and stereotyped form that they can easily be indexed by computer. The *Science Citation Index* is a useful tool for information retrieval. Suppose for example that one has already found a paper, published a few years back, with information on a topic in which one is interested. By looking up all the more recent papers citing this paper, one may perhaps come across further information on the same subject.

This is obviously a perfectly legitimate procedure, justifiable by its practical utility for information retrieval. What is more questionable is whether the number of citations earned by a particular paper is a valid measure of its scientific importance. Although there is obviously a significant correlation between the rate at which a piece of research is cited by other scientists and its influence on the advancement of knowledge, this is not a perfect indicator of the ability of its author or his or her suitability for, say, an academic appointment (§5.5).

Again, when two papers cite the same earlier paper, some sort of link may be inferred between them. These *co-citation* links can be analysed statistically into more or less distinct clusters, indicating that certain groups of scientists interact much more closely with one another than they do with 'outsiders'. These clusters can often be related to the intellectual interests that they have in common, such as the development of a new theory or the solution of a challenging problem (§5.4). The question remains, however, whether this sort of formal analysis reveals significant 'taxonomic' features of scientific knowledge, or of scientific sociability, which would not be apparent to more qualitative inspection. It is one thing to draw attention to citations as general evidence of the 'connectedness' of the scientific enterprise: it is another matter altogether to assume that they exactly mirror the intellectual or interpersonal relationships within or between scientific 'specialties' (§5.3).

4.3 What does a scientific paper say?

On the face of it, a scientific paper is a straightforward report of an investigation designed to answer a specific scientific question (§2.14), and now satisfactorily completed. But this impression is misleading. A scientific paper seldom presents an historical account of what actually happened from day to day in the laboratory. It seldom presents an unbiassed assessment of the state of knowledge prior to undertaking the research, nor of the implications of the results that have been obtained. It seldom even contains a full record of the basic observational results in a form suitable for consultation by a non-specialist (cf. §4.1).

From the author's point of view, a scientific paper is more than a report of work done: it has a very significant *rhetorical* purpose. It is designed to persuade other scientists that its discovery claims are valid, or at least very plausible, and that it can therefore take its place in the archives as a potential contribution to the future consensus on the subject (§1.5).

These intentions are evident in the conventional format and style of all primary scientific communications. A paper must be written as if it were addressed to a hypothetical, very sceptical reader, who is already very well informed on the subject, and might therefore form the spearhead of critical opposition. It is couched in formal technical language, thus indicating the professional competence of the author. Other research bearing on the subject is religiously cited, both to authenticate the basic premises of the investigation and to indicate that the author is thoroughly familiar with all the background material. The theoretical arguments and experimental results are expressed impersonally, in the passive voice, as if to emphasize the objectivity (§3.2) and disinterestedness (§6.2) with which the research was undertaken, and the conclusions are given quasi-logical weight, suggesting the rational necessity (§3.8) of this particular outcome. In other words, a scientific paper tries to carry conviction by apparently reporting an investigation that was conceived and carried out in accordance with the established principles of the scientific 'method'.

This fiction is perfectly well understood by scientists themselves. The fact that a scientific paper is a pious 'fraud' does not mean that scientists are liars and hypocrites. In the actual course of their research, they could never be as farsighted, objective, impersonal, rational, or technically scrupulous as they later make out, since these virtues would not be compatible with the personal commitment and sense of conviction needed to pursue an investigation into previously unknown territory. But when, eventually, they 'write up' their results, it is important that they should anticipate potential criticism by applying these principles to the account of their findings. The pseudo-historical format and pseudo-impersonal style of a scientific paper is the most effective medium by which these findings can then be brought to the attention of other scientists through the narrow channels of the formal communication system. From an epistemological point of view, this is an essential step in the production of testable scientific generalizations (§2.3) and explanations (§2.9) from the observation of particular events and experimental phenomena.

4.4 How do scientific papers get published?

Ideally speaking, all scientists should use the scientific 'method' and produce fully convincing research reports on problems of scientific importance. In reality, scientific papers vary enormously in quality. Of the innumerable primary papers that are published, only a small proportion prove eventually to have made a significant

4.4 How do scientific papers get published?

contribution to knowledge. Many are so uninteresting or so unconvincing that they are never even cited (except, of course, by their authors!). Others are soon shown to be invalid, through relatively elementary errors of fact, method, or principle which should have been corrected by their original authors. Allowance must be made, of course, for preliminary explorations of new fields, reports of inexplicable anomalies (§7.3), theoretical conjectures (§2.13) and other communications suggesting possible fruitful lines of research. Nevertheless, after making every allowance for the wisdom of hindsight, a great deal of the primary literature of science can be seen to have been of very little interest or competence even when it was first published. The fact that an investigation into a scientific question was carried out by a professional scientist, using scientific instruments, is no guarantee that its conclusions are of any scientific value.

In practice, scientific journals do not even publish all the papers that are submitted to them. A substantial proportion of the typescripts they receive are rejected altogether, whilst others often have to be thoroughly revised before they are accepted for publication. The rejection rate varies considerably from journal to journal and from discipline to discipline, but is seldom less than 20%, and may be as high as 80% for a particularly prestigious periodical such as *Nature* or *Science*.

This is a very significant feature of academic science. It must be emphasized that the papers that are rejected are not all obviously crazy, or cranky, or pathetically inept: even a well-formulated paper by an established professional scientist can be at risk in this selection process. The effort needed to get a paper published produces one of the main conflicts between individual and communal interests of scientists. The author's interest is obvious. Since 'productivity' is a crude measure of one's scientific work and achievement, one naturally tries to get as much of one's research published as possible. But this is opposed by the interests of all other scientists as 'readers'. They may not wish to deny this right to a professional competitor, but they do want to maximize the validity, and minimize the number, of the papers they have to read.

Although the communication network of science is not a formally closed system, the number of 'reputable' journals – i.e. the journals scientists normally attend to in their research – is limited, and there is strong competition amongst scientific authors for access to this finite resource. The journals themselves are normally owned by independent bodies, such as learned societies (§7.2) and commercial firms, who have to meet their printing and publishing costs from the income they get from subscribers. In many cases, nowadays, these subscribers are not individual scientists but the libraries of institutions such as universities and industrial laboratories, but that does not lessen the economic pressures on the primary communication system as a whole. The editor of a scientific journal therefore has the intellectual responsibility, the commercial incentive and the legal authority to select for publication those papers that the subscribers to the journal will want to read. An effort must be made to publish papers that are novel, convincing, and relevant to the specialty that the journal serves (§5.3), and therefore to reject papers that appear to be 'unoriginal', 'unsound' or 'irrelevant' to that specialty.

4.5 Selection by peer review

The editor of a small, highly specialized learned journal may exercise his or her own judgement on the typescripts sent in for publication. But few individual editors can have the professional expertise to make a fair assessment of the scientific quality of hundreds of papers each year. The editor (or editorial committee) of a major scientific journal is bound to seek specialist advice on whether or not a particular paper should be published as it stands, or revised in detail, or rejected out of hand. It has become customary, therefore, to consult *referees* – or, in the American terminology, *reviewers* – with detailed knowledge of the topic of the paper in question.

But who could be such a referee, except another scientist working in the same field of research as the author or authors of the paper being assessed – in other words, an actual or potential colleague or competitor of the author? In effect, the decision to publish a research report is put into the hands of one or more of the scientific 'peers' (i.e. social equals) of the person who did the research. The public achievements and reputation of every academic scientist thus depend upon the opinions of the other scientists within the same narrow specialty.

The *peer review* process is evidently a highly reflexive and convoluted social activity, where a delicate balance has to be achieved between three distinct interests – those of the author, of the editor, and of the referee. It may be that this balance can only be maintained because any professional scientist of some standing may be called on from time to time to play any of these roles: it is as if every citizen must sometimes be the accused, sometimes the judge, and sometimes in the jury in a succession of criminal trials! It is not surprising that this process is the focus of considerable contention, both amongst scientists themselves and amongst sociologists of science.

It is often argued, for example, that referees should not be anonymous (as is customary), but that their names should be disclosed to the author, or even published with the paper. Another proposal is that the names of authors should be temporarily removed from typescripts before they are sent out to referees for comment. There is already wide variation from journal to journal on the number of referees that should be consulted, and the procedures to be adopted if they disagree. In the journal of a learned society, for example, should members have a right of appeal to some higher body, such as the council of the society, if their papers are rejected by an editorial board? Then there are niceties of editorial policy, such as whether special care should be taken not to choose referees who are known to be in an opposing scientific camp to the author – or even in the same camp. Is the ideal referee an established scientist of long experience and 'mature judgement' or a relative beginner with closer knowledge of the research front? Should the style and format of a paper be given much weight in the assessment, or should the referee concentrate primarily on its scientific content. Above all, should editors encourage referees to take a severe line, so as to 'keep out nonsense', or should they be lenient in their judgements, in order that possibly valuable observations or insights should not be lost to view.

There is no space here to present the pros and cons in these debates, nor even to report the results of some interesting empirical studies of the peer review system in action. Nevertheless, this is a very sensitive area in the workings of academic science, and a very fruitful topic for metascientific research, since the action here is in all three dimensions of science at once – personal, communal, and intellectual.

4.6 The accreditation process

Very few primary papers are epistemologically self-contained. The validation of a significant discovery (chapter 3) is a long process that can seldom be completed satisfactorily in a single investigation. Indeed, the essence of our model of academic science (§1.5) is that the *accreditation* of knowledge is a *social* process, in which the research claims of every scientist are subjected to critical reassessment by other scientists before they can be regarded as 'well-established' (§3.8). The communication system of science plays a vital part in this process.

The publication of a paper after peer review does not completely accredit its contents. The referee cannot be expected to validate a discovery claim according to strict philosophical standards of proof. A referee's report can be no more than a 'first reading', certifying that the material is original and not trivial, that the previous literature has been taken into account and cited, that the argument is clearly expressed and not implausible, that the experimental procedures are technically competent, and that the conclusions are not contrary to indisputable facts. A good referee's opinion may be intuitively sound but it can only be based upon a superficial inspection of the author's typescript, without access to the actual research apparatus, laboratory notes, computer programmes, etc., out of which the paper was produced.

As a consequence, the information published in the primary scientific literature must be treated with some caution. Primary papers often contain factual errors or theoretical inconsistencies which have not been noticed by their authors or picked up by referees. More seriously (§4.4), only a small fraction of what is initially claimed turns out to be so well founded and significant that it is eventually incorporated in the general scientific consensus. The further stages in the accreditation process are more diverse and not so clearly defined as acceptance for publication, and usually take many years.

Thus, a paper may be cited unfavourably in another primary paper, on the grounds that it contains errors, or is made obsolete by a more thorough investigation. Needless to say, this is often the first shot in a scientific *controversy*, in which different groups of scientists make claims and counter claims concerning the solution to some specific problem (§2.14). Such competition for accreditation is usually very beneficial to science, since it ensures that the answer that eventually gains acceptance has been exhaustively tested by hostile critics.

A favourable citation adds something to the credibility of a primary paper – but what is one to make of it when it is not cited at all in a subsequent publication on the same subject? Is this due to ignorance, or does it imply that the earlier paper was thought to be so unconvincing, or so trivial, that it was not worth mentioning? It is not the bounden duty of a primary author to review the whole past literature of the subject before reporting the results of a new investigation, so that it is often wiser to 'forget' a weak contribution than to point out its deficiencies. Nevertheless, in the diplomatic language of scientific discourse, silence speaks against assent.

The formal responsibility for putting all the primary contributions to a subject into some sort of order rests with the authors of *review articles*, *research monographs* and similar 'secondary' publications. Ideally, this should be an objective, unbiassed survey of the evidence, putting the case for and against each of several conflicting views and leaving it to the reader to decide between them. But such a survey is practically useless unless it conveys the author's opinion as to which discovery claims can now be regarded as well established, or at least worth pursuing, and which are probably now best 'forgotten'. Needless to say, this is often a very delicate task, calling for considerable tact as well as good scientific judgement.

An 'important' discovery that has successfully passed this stage of public validation may now be a candidate for more positive accreditation, such as the award of a *prize* (§5.1). But it must not be assumed that this makes it quite incontrovertible. A significant proportion of the supposed discoveries that have earned high recognition for their authors, such as election to a National Academy, or even a Nobel Prize, have not survived as permanent features on the scientific map. On the other hand, this stage of positive accreditation may be long delayed, even for a very important discovery such as the Theory of Continental Drift, because it is still publicly opposed by a few influential scientists (§7.3).

The accreditation process continues into the 'tertiary' literature of science, such as encyclopaedia articles and *textbooks*, which do not pretend to cite all the primary literature, but merely give a few references in support of the major factual discoveries or widely accepted theories that are selected for exposition. One of the most potent media for disseminating the current consensus in an academic scientific discipline is a course of undergraduate *lectures*, in which research results of 20 or 30 years ago are usually presented as if they were now beyond question. This dogmatism is not always justified – but given another 20 years or so, some of this material will be incorporated into school science syllabuses (§16.2), where, of course, it is entirely immune from criticism.

4.7 'Informal' communication between scientists

This chapter has concentrated on the 'formal' communications carried out by journals, books, and other archival publications. But any description of scientific work would be incomplete without reference to the 'informal' transfer of information between scientists, in face-to-face conversation, telephone calls, lectures, letters, the exchange of 'pre-prints' of papers and so on. This activity is not systematized, although it is often facilitated by semi-formal social occasions, such as seminars, symposia, conferences and other scientific *meetings*, which may give rise to archival publications such as *conference proceedings*.

These informal channels of communication obviously play a vital part in the research enterprise (§5.4). They are the means by which speculative ideas, technical wheezes and other 'unpublishable' items of information diffuse through the scientific community. They are often the means by which significant new developments come to the knowledge of active research workers, long before they can be formally published. At the most immediate level of personal interaction, scientists usually spend a great deal of time simply talking 'shop', mulling over ideas, discussing possible interpretations, and generally stimulating each other to thought and action. As scientists in developing countries well know, it is very difficult to carry out research in isolation, without personal contact with scientific colleagues.

But how should this sociability be dealt with at a metascientific level? The traditional view seems to be that the channels of the informal communication network simply run in parallel with the formal channels, speeding up the transfer of information throughout the system. The internal social and intellectual structure of science would thus be represented in broad outline by the formal links, even though the 'strengths' of these links might be significantly modified by hidden informal connections. Thus, the sociologist or historian of science would start from, say, a citation mapping of the archival literature (§4.2), and then look for other influences such as private conversations and correspondence.

Some sociologists, however, go further, and treat the *informal* network as the dominant mode of communications between scientists. According to this view, the formal literature seems little more than an epiphenomenon, generated by the ritual activity of 'writing up the results, for the record'. In other words, scientific knowledge can no longer be securely located in its archival publications, but consists mainly of 'what scientists know, and talk about amongst themselves'. To understand 'laboratory life' properly, one should therefore make 'ethnographic' studies of research in progress (§1.4).

The results of such studies are at first somewhat disconcerting, for they show that scientists are singularly unconstrained by the principles of the 'scientific method' in their daily work. If this observation is taken at its face value, it has profound consequences for epistemology. The notions of scientific objectivity and validity seem to dissolve, and the way is open to a sociology of knowledge (chapter 8) in which science seems so like any other type of social activity that it can claim no higher credibility than any other body of socially accepted beliefs.

But a narrowly ethnographic approach tends to underestimate the considerations that motivate research workers on a longer time-scale than the day-to-day manipulation of their apparatus. One cannot account for the behaviour of scientists without reference to their own widespread belief that the public communication of their research results, in a form that must be intellectually acceptable to other scientists, is an essential part of being a scientist at all – or at least an 'academic' scientist, as is assumed throughout the present chapter (§6.4). As we have already remarked (§4.3), it is precisely the process of 'writing up' that transforms the contingencies and expediencies of the laboratory into a contribution to science. Since every scientist knows that published papers are the only demonstrable output of research, the knowledge that such papers must in due course be produced has an imponderable but pervasive influence on the course of laboratory life. These influences thus put the ethnographic model back towards the more conventional scheme in which the formal communication system plays a major role.

It is quite true, all the same, that the public scientific archives, primary, secondary and tertiary, seldom give a clear response to the question 'What does science know about X?' The primary literature may be confused and contradictory, the secondary literature equivocal, and the tertiary literature dogmatic but out of date. The communication system is not, in fact, designed to give direct answers to most questions of this kind. Much that is known is stated only by implication, or can be tacitly assumed (§3.3), or is only really understood by scientists themselves, as a consequence of their research experience. Thus, the notion of a 'scientific consensus' on a particular point (§8.5) is ill-defined, and should be considered an ideal, or a goal, rather than an achieved reality. Science is communicated publicly, by formal or informal means, in order to grow and change itself. Scientific communication is a *dynamic* process, with *evolutionary* epistemological consequences (§3.8), and has little to do with the statement, or restatement, of static positions already arrived at.

Further reading for chapter 4

A good elementary text, with references to most areas of current research or controversy, is

    A. J. Meadows, *Communication in Science*. London: Butterworth, 1974

A partisan, but honest introduction to a powerful methodology is given by

    E. Garfield, *Citation Indexing*. New York: Wiley, 1979

A fascinating paper, with accompanying commentaries covering almost every controversial aspect of the peer review process is

    D. P. Peters & S. J. Ceci, 'Peer-review practices of psychological journals: the fate of published articles submitted again'. In *The Brain and Behavioural Sciences*, **5**, 185–255 (1982). Republished as *Peer Commentary on Peer Review*, ed. S. Harnad, Cambridge: Cambridge University Press, 1982

The way in which the writing of a scientific paper transforms the events it supposedly reports is shown by

    K. D. Knorr-Cetina, *The Manufacture of Knowledge*. Oxford: Pergamon, 1981 (pp. 94–135)

The rhetorical features of scientific papers are discussed by

    M. Mulkay, J. Potter & S. Yearley, 'Why an Analysis of Scientific Discourse is needed'. In *Science Observed*, ed. K. D. Knorr-Cetina & M. Mulkay, pp. 171–204. London: Sage, 1983

5
Authority

'To punish me for my contempt for authority, fate made me an authority myself.'

Albert Einstein

5.1 Recognition

Scientists make 'contributions' to knowledge: what do they get in return? Nowadays most scientists are paid a salary to do research, either on a full-time basis or as a normal part of their academic duties (§10.4). From the point of view of an economist, they are simply professional employees, earning a living by their labour. A psychologist, on the other hand, might emphasize the peculiar personal gratifications of research and discovery, for which there is ample testimony in the autobiographical writings of (mostly successful) scientists. In practice, scientists (like other people) respond to a complex mixture of professional and vocational incentives, which arise from the social environment in which they live and work. Sociologically speaking, academic scientists get both their psychological and material inducements primarily through membership of the *community* of other scientists. Satisfactory research performance earns *recognition* within the scientific community, which is usually linked to more obvious rewards from society at large.

Scientific recognition takes a variety of forms, graded to the various stages of a successful career. At the very lowest level, an academic scientist scarcely exists unless his or her work has been *published* in a reputable scientific journal (§4.4). But a published paper is of little significance unless it is *cited* in papers by other scientists (§4.2). Even though citations are not always favourable, they do indicate that the cited author is doing research that is worthy of some notice. The next level is more subtle, being signalled by the *attribution* of some phenomenon or concept to a particular researcher, indicating that it is likely to prove a permanent item of established knowledge (§3.8). Just as 'imitation is the sincerest form of flattery', so the most genuine act of scientific recognition is to treat a discovery as valid, and to build further work upon it. This is often recognized *eponymously*, by attaching to it the name of the discoverer, as in *Addison's* disease, *Banksia*, *Einstein's* equation,

5.1 Recognition

etc. – although where there are several authors, or where several scientists have made the same discovery simultaneously, they may have to share the honours in a barbarous compound label, as in the *Epstein–Barr* virus, or the *Wentzel–Kramers–Brillouin–Jeffreys* (WKBJ, for short) approximation.

These indications of recognition as a scientist arise within the normal procedures of the system of communication. The formal institutions of the academic community (§7.2) recognize individual scientific achievements more directly and personally, by *honorific* awards, such as prizes, medals, honorary degrees, or membership of an élite body such as the Royal Society of London or the United States National Academy of Sciences. In the modern world, the winners of Nobel Prizes are selected annually by the Royal Swedish Academy of Sciences and its medical counterpart. The members of this international ultra-élite of the sciences are only the most highly acclaimed of the several thousand researchers in various countries who have received some sort of formal acknowledgement of their professional standing amongst their fellows.

Active participation in the everyday business of the learned world implies recognition as a member of that world. At the lowest level, the right to take part in a scientific conference (§4.7) by 'reading a paper' indicates some degree of competence to contribute to the proceedings. An unsolicited invitation to chair a conference session or to give a review lecture, although superficially a part of the process by which knowledge is communicated and accredited (§4.6), is highly indicative of recognition as an *authority* on the subject in question. The same message is conveyed, in private, by a confidential request to act as a referee for a paper submitted to a learned journal (§4.5) or for an application for a research grant (§14.4). These purely scholarly roles shade into the more practical responsibilities of journal editors, and 'honorary' officers of learned societies, who are often chosen for their scientific standing rather than for their managerial capabilities.

Communal recognition may also be indicated by the material rewards of academic or other institutional office. Thus the early stages of scientific *employment* are largely dependent upon the presentation of a formal *curriculum vitae* in which the candidate's list of publications is the major item (§10.4). Further steps up the academic ladder, including such vital steps as getting permanent *tenure* of an academic post, are similarly linked to recognizable scientific achievements, which they in turn publicly accredit. There is obviously a very strong material interest in getting, and keeping, a well-paid job, but the symbolic significance of personal preferment should not be underestimated. A scientist who seeks an appointment as a Full Professor may be less concerned with the extra money that he will get than with a sign of successful performance as a research scientist. This sign is clearly visible to the outside world, perhaps leading to further external recognition as an expert (§15.5) on government committees, in the media, or in industry and commerce.

At a certain stage in this upward progression, however, the material rewards, the powers, and the responsibilities of managerial or administrative authority become disconnected from personal contributions to science. A 'recognized' scientist who becomes the head of a university, or a government agency or a large laboratory, has, so to speak, left the scientific community and must look for career standards, goals and incentives in the larger world (§14.6). When Robert Oppenheimer took leave from his position as Professor of Physics at the University of California to become Director of the Los Alamos Laboratory, he was entering a different stratum of society, where managerial and political skills took precedence over research ability and achievement. Nevertheless, as we shall see in later chapters (e.g. §12.4), such skills can prove of the greatest importance within the scientific enterprise.

5.2 Exchange of gifts – or competition?

Some sociologists have suggested that the scientific community functions by the *exchange* of 'communications' for 'recognition'. This model of social activity is derived from social anthropology, where there are many examples of elaborate ceremonies in which valuable objects are given 'freely' by members of a community to other members, in the confident expectation that they will eventually be repaid in material or symbolic kind. According to this model, there is nothing remarkable in the fact that scientists do not insist upon immediate cash payments for their 'contributions' to knowledge, but rely upon the community to provide later pay-offs in material goods or social esteem.

The analogy between a scientific conference and a potlatch ceremony is instructive, but does not provide a complete explanation of why scientists behave as they do. In particular, it does not take account of the fact that 'recognition' is itself a social construct, which is always in short supply. As everybody knows, academic science is characterized by fierce *competition* for the limited number of honorific awards and senior posts that can be won by research. Scientists who work hard throughout their careers, and produce a respectable number of communications, are often rewarded very meagrely for their efforts.

The exchange model is built upon the assumption that most scientists will be seen to get a fair deal from the community. This assumption can only be maintained if it is also accepted that there is a very large variation in the relative *quality* of their contributions. As Einstein 'deserves' a Nobel Prize for a few brief communications because one of these was a really 'important' discovery: Bloggs will have to be satisfied to retire as an Associate Professor because the 65 papers he has published are all rather 'trivial'.

By any objective measure, scientific communications vary greatly in the value of their contributions to knowledge (§4.4), but there is no single measure by which

5.2 Exchange of gifts – or competition?

they can fairly be ranked for recognition. The apparent significance of a discovery claim not only changes with time as new information becomes available (§3.8): it is also strongly dependent upon *opinion* within the scientific community concerning the validity of theories and observations to which it relates (§7.5). In other words, the 'quality' of a communication is not an entirely empty concept, but it can be just as much a social construct as the degree of recognition for which it is exchanged. It is salutary to recall, for example, that Einstein did not get his Nobel Prize for his famous paper on relativity theory, which was still regarded with suspicion by some physicists, but for his work on the photoelectric effect, which was acceptable to everybody although in some respects less profound.

For this reason, it is difficult not to fall into a circular argument, in which the quality of a communication is measured by the recognition it receives. The competition for personal recognition is inseparable from the process by which scientific discoveries are continually being assessed and reassessed. In principle, this is critical evaluation by objective validation (§3.8) and communal accreditation (§4.6): in practice, it is so closely linked to personal and group interests that social *negotiation* might be a better description of what goes on.

Personal competition for communal recognition drives and shapes academic science. It is a powerful incentive to individual discovery and to collective criticism, but it does not always work in the best interests of scientists or of science, for it tends to undervalue the cooperative attitudes and activities that are also essential to the research enterprise. Thus, for example, the conventional criteria for an original communication (§4.2) simply do not 'recognize' work such as managing research facilities, or writing textbooks, which may call for as much scientific talent as making discoveries. In some cases, the credit for an important scientific advance should really go to an 'informal' communication, such as a comment in conversation, which never gets acknowledged in the formal literature. Where official secrecy is imposed, quite pathological situations can arise: it is ironical that Andrei Sakharov, now the dissident champion of openness and cooperation in science, was made an Academician, and thus received public scientific recognition, for allegedly brilliant work on the Soviet H-bomb which has never been published.

The whole tendency to attribute each major scientific achievement to a single individual leads to notorious injustices. It makes insufficient allowance for team research (§11.4), for simultaneous discovery, and for the cumulative effect of a succession of interconnected researches by many different people, leading step-by-step to the solution of a long-standing problem. A careful account of any major scientific development, such as the elucidation of the biological function and chemical structure of DNA, will show the true human complexity of motive and action within the discovery process.

5.3 Specialization

Research is a demanding career. High levels of individual performance are called for to earn recognition. The standards of originality and significance required of a publishable communication (§4.4) are seldom reached without personal effort. Success against competition can only be achieved by extreme *specialization*. To gain and keep a place in the scientific community, it is practically essential to concentrate one's research on a very narrow range of problems in a very restricted field within a particular discipline.

The high degree of specialization in science is a familiar fact that is often deplored; but it follows naturally from the logic of the situation of every scientist. The academic scientist must know the literature of a subject well enough to be aware of significant research problems (§2.14) that have not yet been solved and to formulate plans to investigate them. It is simply impossible to become acquainted with the contents of more than, say 1% of the 100000 or so primary papers that are published annually in a major scientific discipline (§4.1), so these must be selected in a correspondingly narrow range of topics. Any particular investigation calls for knowledge and skills that can only be acquired by experience of research on the similar problems to which it will be linked by citation in the subsequent report (§4.2). The whole philosophy of the scientific enterprise is based upon the accumulation of facts and theories that need be validated only within the contexts to which they apparently relate (§3.8), thus permitting a very fine-grained intellectual division of the labour of research.

This differentiation into intellectual *specialties* is closely associated with and reinforced by various forms of communal differentiation. The lower stages of the process of recognition (§4.1) are limited to small domains on the 'map' of a discipline. A researcher requires a certain standing in a particular field to be entrusted with the facilities, such as apparatus and assistants, to undertake independent research in that field (§11.2), but this might not be sufficient for him to gain a grant for research (§14.4) in another field to which he or she had not previously contributed. A scientist who has gradually acquired a reputation by the accumulation of contributions to a particular specialty may remain practically unknown in other fields of the same discipline or sub-discipline. Indeed, quite severe difficulties arise when there is direct competition for an honorific award between scientists whose research has been confined to entirely distinct specialties – as it might be, for example, in the comparison between a radioastronomer and a metallurgist for a Nobel Prize in Physics. The whole machinery of assessment, accreditation and recognition is mainly designed to work 'locally'. Attention is focused on citations, skills, contributions, etc., that relate to a particular scientific problem, with little reference to achievements or competences that might have more general applicability.

5.4 Invisible colleges

Scientists mainly interact communally with other scientists in their specialty – that is, with other members of the *invisible college* of their field of research. This is not, of course, a precisely defined group, since it consists simply of the research scientists who happen at the time to be trying to solve a particular scientific problem, such as the origin of the planets, or who are using a particular experimental technique, such as electron microscopy, or who are interested in some particular aspect of nature, such as the growth of plants. It is not institutionally, geographically or nationally localized: when I was myself doing research on the electron theory of liquid metals, I counted amongst my colleagues not only several British physicists, but a number of Americans, several Japanese, and others from a variety of countries including India, France, Israel, Canada, Sweden and Italy.

Most of the activities of an invisible college are informal and unstructured (§4.7). The members communicate with one another by letter and telephone, send each other pre-prints and reprints of their papers, arrange and attend conferences and summer schools on subjects of common interest, visit one another's laboratories for shorter or longer periods, and give post-doctoral posts to each other's students. These social interactions are not institutionalized, but draw their significance from the corresponding linkages in the cognitive domain. Thus, an invisible college has its counterpart in the scientific literature, as a closely connected cluster of nodes in a co-citation network of scientific authors (§4.2). Although these clusters are seldom sharply differentiated, and often overlap in complicated ways, they are quite genuine groupings whose connections are usually mirrored in the social interactions of their members. In many cases, a high proportion of the papers linked in a co-citation cluster would have been published in the same few journals (§4.4): the formal communication system of science is strongly differentiated and specialized to meet the narrowly defined interests of groups of authors and readers. This, in turn, generates closer social interactions within the invisible college, whose members may have to work together on editorial boards, peer-review panels, academic appointment committees, and other more organized collectives. Indeed, the college may eventually become 'visible' as a small learned society or professional institution (§7.2).

From the point of view of the individual scientist, the 'scientific community' is as vast and abstract as a whole nation: the village of scientific life is the invisible college of a specialty. It is within this microcosm that recognition is sought and reputation won. This is the little world of a few hundred people to which the graduate student is socialized, where rôle models are discovered (§15.2), where esteem is so much desired, and within which most research careers are effectively contained. The large-scale official institutions of academic science, such as the national academies and professional institutions, are much weaker, and much less compelling, than the unofficial specialty groupings where most of the action really occurs. In this respect, also, the internal social structure of science reflects its cognitive structure, which is loosely articulated and incoherent over all (§3.8).

5.5 Stratification

The scientific community is far from egalitarian. The successive stages of recognition (§5.1) give rise to a distinctive social *stratification* within science. In the competition for esteem and authority (§5.2) only a small proportion get the prizes. Thus, for example, not all those who get tenured posts in academia rise to the rank of full professor; only a few per cent of science professors become Fellows of the Royal Society; of nearly 900 Fellows of the Royal Society, only 30 or so are Nobel laureates. Although members of the higher strata do not necessarily have much more organizational power than those of lower status (Nobel laureates cannot give *orders* to other professors), they may exercise considerable influence through the informal relationships of the scientific community. This stratification of esteem thus parallels an informal, but tangible, social structure within the world of academic science.

Scientific *authority* is exerted through a variety of channels. 'Senior' scientists are influential in assigning recognition to those more junior: it is they who take the decisions, or who are consulted, about publications, promotions, and prizes. They have power over the allocation of resources for research, through such academic posts as Departmental Chairman or Dean of Faculty, or indirectly as referees and members of review panels of funding agencies (§14.4). They also speak for the scientific community in its 'external' relationships, whether on technical issues or on general policy matters (§10.5, §15.4).

Almost inevitably, even if unconsciously, this influence is used to their own advantage within the scientific community. Prestige in science is self-reinforcing. A young scientist who makes some recognizable contribution in a PhD dissertation gets a post-doctoral appointment at an élite institution, such as Cambridge or Harvard or the Tata Institute in Bombay. Association with, and visibility to, the local notables helps in the competition for a permanent post at the same institution. This, in turn, provides access to brighter students and colleagues for collaboration (§11.4) in further research projects. The scale and competence of this work enhances the reputations of those involved, and gives it priority in the allocation of resources (§14.4), thus opening the way to further significant investigations. Surrounded by an active research group, recognized as a core member of an invisible college, chairing a grant allocation committee, placing past colleagues and students in other strategic posts, esteemed publicly as a 'leading authority', and distributing minor honours to others, such a person has become essentially impregnable to criticism: even the scientific

5.5 Stratification

papers to which he or she puts his or her name may be reviewed much less stringently (§4.5) than if they were by a relatively unknown, more junior scientist.

Scientific authority develops under the influence of the *Matthew effect*: 'For unto every one that hath shall be given, and he shall have abundance: but from him that hath not shall be taken away even that which he hath', as it is put in the *Gospel according to St Matthew*. The general principle of *cumulative advantage*, which makes the rich grow richer, and the poor grow poorer, clearly works within academic science as it does in any stratified social system. This does not necessarily mean that scientific authority is acquired illegitimately or exercised corruptly. But it does raise the question whether the social stratification of the scientific community is a reflection of genuine differences in some individual variable, – some intrinsic personal quality that could be called, say, *merit* – or whether it is just a case of some people being lucky at an early stage in their careers, and then rising to the top by the Matthew effect alone.

It is so difficult to disentangle personal abilities from the social context in which they developed that this question cannot be answered convincingly. Nevertheless, scientists themselves are largely agreed that scientific 'talent' is a personal characteristic that is very unevenly distributed, even amongst professional researchers. It has often been argued, for example, that this variable should be expressed on a logarithmic scale: according to a famous remark attributed to L. D. Landau, 'a "Grade 1" scientist would be one who has contributed 10 times as much to knowledge as one in "Grade 2", and so on'. It is also argued that the number of scientists increases by a large factor from grade to grade, so that mediocre scientists far outnumber those with genuine talent.

This highly subjective opinion is supported by more objective evidence, such as crude measures of individual scientific *productivity*. Thus, scientists differ enormously in the number of papers they publish. According to *Lotka's Law*, the number of people who publish n papers, or more, is proportional to $1/n^2$. For example, about 10% of all scientists have published at least 10 papers, whereas only 0.1% have published 100 papers or more. This formula is not to be relied on in detail, but it does give some idea of the extreme skewness of the distribution of this particular variable. It may be insisted that this sort of productivity has little to do with real scientific merit; but detailed counts of *citations* (§4.2) similarly indicate that a small proportion of all published papers, by just a few authors, get the bulk of subsequent attention by the scientific world.

It does seem, therefore, that there is as high a degree of stratification in the supposed talents, and in the objective achievements, of research scientists, as in the authority they are accorded in the scientific community. But that does not prove that merit is always rewarded with success, or that the best scientists must be those at the top. The mechanisms of self-fulfilling prophecy, of which the Matthew effect is an example, confound any deeper analysis of such issues.

5.6 Functions and dysfunctions of authority

An effective scientific community is an essential feature of academic science. This community is often treated by sociologists of knowledge as no more than a cultural context, in which the current scientific consensus is articulated and sedimented, and tacit skills and intellectual traditions are transmitted from generation to generation of scholars (chapter 8). But it provides other collective resources and functions which are absolutely necessary for its individual members. Some of these functions, such as the institutionalization of an efficient system of communication (chapter 4) and of an equitable procedure for the allocation of research facilities (chapter 14), are fairly practical and straightforward. Others, such as the regulation of competition for recognition, are highly symbolic and convoluted.

The peculiarity of the internal social system of science is that it is highly *stratified*, but not hierarchically *structured*. Responsibility for the performance of these important functions thus rests upon the members of a loosely articulated *élite* whose status derives mainly from their personal scientific achievements. Although they sometimes exercise formal authority, such as editing a journal, presiding over a learned society or chairing an appointment committee, they usually have little direct power as individuals. The very real influences they exert are charismatic, through their supposed intellectual status; paternalistic, through the patronage of jobs and prizes; and oligarchical, through innumerable committees, councils, boards and panels. Primarily located in their separate specialties (§4.3), dominant in the leading circles of their various invisible colleges (§4.4), autocratic in their own academic departments, they nevertheless constitute a landed aristocracy in a country without a parliament or a king.

This social system is more or less consistent with the regulative principles of academic science (§3.10) and with its epistemic method. But can it cope with the increasing pressures of internal cognitive change and external demand? Is it compatible, for example, with the provision and use of very large-scale research facilities, such as particle accelerators and space probes, which have to be managed by bureaucratic methods (§11.5)? The superstar scientist who makes a great discovery may well be a very intelligent and high-minded person, but those qualities are not all that is needed to take charge of an organization employing thousands of people. A talent for formulating new theoretical concepts is good for assessing the quality of other people's scientific papers, but it does not necessarily transfer to the scrutiny of a £10 million budget, or the precise wording of a political document. If the internal social system of science has to become more administratively efficient to meet

5.6 Functions and dysfunctions of authority

these pressures, then the membership and rôles of the various strata of authority also have to change. This is a theme which will be taken up again in chapter 12.

The traditional structure of academic science is also notably undemocratic. Recognition always comes by nomination from above, rather than by election from below. The constitutional procedures of the older scientific institutions are incredibly oligarchical: in the Royal Society, for example, executive action is decided by the Council, which effectively names its successors and practically never consults the Fellows on general policy issues. Such procedures may be acceptable in an organization whose business is not of great public significance, but they are very unresponsive to voices of dissent and forces for change. Now that research is a major profession and science is a major institution in society, is there a need for more open communal organizations where the hopes and fears of the tens of thousands of 'junior' research scientists can be expressed and attended to?

For it must always be remembered that 'authority' is ambivalent in science. It is always wise to attend to its advice, but a folly to follow this without question. The power to patronize a promising talent can easily be corrupted by flattery, and the gratification of public recognition for one's work is sadly conducive to excessive vanity. These dysfunctions of authority can be kept in check by open discussion within the community at large, and by competitive individualism within the élite itself. But when these checks fail, there is nothing to prevent the top strata from consolidating themselves into a self-perpetuating and self-regarding *Establishment*. It is a matter for serious consideration, in the sociology of science, whether this actually occurs. Is there a coherent ruling group in the scientific community that suppresses intellectual innovation, exploits the labours of the mass of scientific workers, and allies itself with the ruling groups of other entrenched classes or social sectors, such as the Military or Industry (§14.6)?

The questions raised in this section have not been convincingly answered or even closely studied. But they need to be considered carefully, for they make manifest our need to have a clearer idea of the internal values and goals of the scientific community, the processes by which scientific knowledge grows and changes, and the relation between science and its external social environment. These topics will be taken up in the next few chapters.

Further reading for chapter 5

Recognition and status are discussed by
    A. J. Meadows, *Communication in Science*. London: Butterworth, 1974 (pp. 172–206)

A general source on the 'superélite' of science is
    H. Zuckerman, *The Scientific Elite: Nobel Laureates in the United States*. New York: The Free Press, 1977

The case for the 'exchange' model, together with a great deal of empirical evidence on attitudes towards competition, cooperation and specialization is presented in

> W. O. Hagstrom, *The Scientific Community*. New York: Basic Books, 1965

Succinct extracts from this classic work are reprinted in

> B. Barnes (ed.), *Sociology of Science*. Harmondsworth: Penguin, 1972 (pp. 105–20)

and in

> B. Barnes & D. Edge (ed.), *Science in Context*. Milton Keynes: Open University Press, 1982 (pp. 21–34)

The growth of specialties is discussed by

> D. Crane, *Invisible Colleges*. Chicago: University of Chicago Press, 1972

Data on differential productivity, and other putative indicators of 'merit' are given by

> D. J. de S. Price, *Little Science, Big Science*. New York: Columbia University Press, 1963 (pp. 33–91)

and by

> H. W. Menard, *Science Growth and Change*. Cambridge, Mass: Harvard University Press, 1971 (pp. 84–128)

A number of important papers on the reward system of academic science are to be found in

> R. K. Merton, *The Sociology of Science*. Chicago: University of Chicago Press, 1973 (especially 'Priorities in Scientific Discovery', pp. 286–324 and 'The Matthew Effect in Science', pp. 439–59)

Empirical evidence on the correlation of 'rank' with 'merit' is presented in

> J. R. Cole & J. S. Cole, *Social Stratification in Science*. Chicago: University of Chicago Press, 1973

For an entry into recent research on these topics, one should look into

> N. Elias, H. Martins & R. Whitley (ed.), *Scientific Establishments and Hierarchies*. Dordrecht: D. Reidel, 1982 (especially the contributions by P. Weingart (pp. 71–81), E. Yoxen (pp. 112–43), J. Fleck (pp. 169–218) and R. Whitley (pp. 313–57))

6

Rules and norms

'I do not know the Game-laws & Patent-laws of science.'   *James Clerk Maxwell*

6.1 Behaving as a scientist

Academic science is not formally organized as a whole. It is not governed by a bureaucratic hierarchy, like an army or an industrial firm (§5.6). It does not have a constitution, a charter, or an official book of regulations. In principle, it is simply a community of individuals, each of whom has a permanent tenure of an academic post as a teacher or researcher. To adopt traditional political metaphors, academic scientists are like free citizens of a democratic republic of learning, or like a community of farmers, each secure on his own holding.

And yet this community is not a mere collection of individuals. Although it does not have an overall organizational plan, it is structured around a number of formal institutions, such as learned societies, and informal institutions, such as invisible colleges. It is spanned by an elaborate communication system which follows standard practices in the management of publications and archives, regulates the rôles of authors, editors, and referees, and has strict conventions on the style and format of papers (§4.5). The procedures by which scientists are 'recognized' (§5.1) are less systematic, but are just as elaborate. For example, intellectual property rights are acknowledged by citation, and there are well-established constraints on the expression of critical comments and other overtly competitive behaviour. The social differentiation of the scientific community into specialties (§5.3) and strata of prestige (§5.5) is further evidence of a complex informal structure, which is sustained by various social regularities, such as the way in which the domain of a specialist is defined, or the amount of deference due to a senior scientific authority.

The freedom of the academic scientist within the republic of learning is thus heavily qualified. Certain types of egocentric or idiosyncratic behaviour would not be consistent with the maintenance of this internal social structure. Some forms of interpersonal behaviour must, in effect, be forbidden, whilst others are permitted or encouraged. If one is to belong to this community, if one is to be recognized as a scientist at all, one must know that there are things that one may do – and things that are simply 'not done'. In other words, a professional scientist must be familiar with the *rules* of scientific behaviour, and must be ready to abide by them in practice.

The fact that these rules are not all formally codified or legally binding does not mean that they are without substance, or of no consequence. Scientists themselves may have difficulty in formulating them precisely, but they usually know pretty well when they have been seriously infringed. Consider, for example, the following questions, relating to recent episodes in the history of science:

Did James Watson use Rosalind Franklin's data, which had been privately communicated to him, without public acknowledgement?

Were the proponents of the Theory of Continental Drift prevented from publishing their ideas in the 1930s, for fear of not getting jobs?

Should the Royal Society have made a public pronouncement on the validity of the Theory of Special Relativity, which was disputed by Herbert Dingle?

Whatever the actual facts about these episodes, their significance can only be appreciated by reference to the way in which scientists are expected to behave to one another in their scientific work.

In these particular cases, of course, the points at issue are very clear. All academic scientists agree that private communications should always be acknowledged, that deviant opinions should never be suppressed, and that a learned society is not bound to express a collective opinion on a scientific issue. But there are much more complicated cases where the correct behaviour is much less certain. This is seldom due to the absence of any relevant rules: the difficulty is usually to reconcile the many conflicting rules that seem applicable. Some rules, for example, such as the prohibition on presenting false data, are obviously far more imperative than, say, the convention that the first person singular should never be used in a scientific communication.

A far-reaching investigation would thus be needed to determine the substance of these rules, their relative priorities when they come into conflict, and the extent to which they vary from one sub-culture of science to another. Such a naturalistic study would also uncover the means by which individuals acquire knowledge of the rules, and how sincerely they actually followed them. It would also inquire into the methods for detecting infringements, and the penalties for non-observance.

Sociological studies of this kind have already thrown light on certain aspects of scientific behaviour, especially in connection with the communication system of science. But empirical social observations are usually incomprehensible without reference to social theory (§2.14). The relatively superficial conventions of personal behaviour in science need a deeper interpretation, in terms of some general theory of social action, before they can be said to make any sense.

One such theory which seems to activate the ethnographic approach to the sociology of science (§1.4), is that the behaviour of scientists is not essentially different from the behaviour of other people in comparable situations. This research programme therefore tends to emphasize the 'ordinary' characteristics of scientific life, such as the tendency of individuals to take a realistic, opportunistic attitude towards social conventions and to treat ideas as bargaining counters in interpersonal negotiations. There is no reason to deny the validity of such investigations, nor of any of the social models around which they are developed – but of course their results are more significant for sociology and social psychology in general than for the study of science as such. It is surely axiomatic that scientists are just 'people' in almost all important respects.

Nevertheless, scientists – like other members of distinct professional groups, such as lawyers, doctors, soldiers and railway drivers – do have specific behaviour patterns that call for further special interpretation. Some sociologists argue that the scientific community is characterized by a general coherent code of *norms*, from which a great deal of this specific behaviour can be deduced.

It would again take the discussion far away into general sociology, social psychology, law and ethics to try to define or explicate the notion of a 'norm'. In any case, this is another highly controversial topic, especially when it is supposed to explain all social behaviour within science or in society as a whole. Our hypothesis here is much more modest. The general idea is that individuals internalize the various norms, and refer to them consciously or unconsciously, in circumstances where established conventions or habitual practices do not seem to apply. Thus, for example, the norm of 'honesty' is not only the principle underlying the existing laws against, say, bank fraud, but can be extended to new situations, such as 'computer fraud', before they come under specific legal prohibition. The social order is thus maintained (at least to some extent) by a framework of principles that are more abstract and general than the specific forms of behaviour that they cover (cf. §2.9).

By applying a scheme of this kind to some aspects of scientific life and work, we are not committing ourselves to a strongly *functionalist* model of the scientific community (§1.5). All that we are suggesting is that the familiar concept of a social rule can be generalized. Just as the ordinary notion of 'honesty' is not vacuous, and is used very effectively to describe, and partially explain, a wide range of ordinary social action, so also there are general terms that explain many of the observed regularities of scientific behaviour. This is at least a reasonable hypothesis, to be adopted provisionally, and tested against the empirical evidence.

6.2 The Mertonian norms

The suggestion that the behaviour of academic scientists could be related to a compact and coherent set of norms was first put forward in 1942 by R. K. Merton. Whether or not it is acceptable in detail, this scheme illustrates the empirical and theoretical issues that arise in this type of sociological analysis of the scientific community. The *Mertonian norms* are not, of course, a precisely defined and standardized code, and are often 'honoured as much in the breach as in the observance'. They may be expressed and exemplified as follows.

Communalism: *Science is public knowledge, freely available to all.* That is to say, the results of research do not belong to individual scientists, but to the world at large. Scientific discoveries should be communicated immediately to the scientific community by publication in the open literature, which anyone may draw upon for their own further use. This norm obviously governs many of the conventions of the traditional communication system of science, such as that academic scientists may not charge fees for citing their work and that, conversely, only published contributions are normally 'recognized' for the award of prizes (§5.1). The legitimacy of this norm is implicitly acknowledged when academic scientists deplore the prevalence of secret research, or where there is dispute over patents or other legal claims to personal scientific 'property'.

Universalism: *There are no privileged sources of scientific knowledge.* That is to say, discovery claims and theoretical arguments should be given weight according to their intrinsic merits, regardless of the nationality, race, religion, class, age – or scientific standing – of the person who produces them. This norm, also, is the guiding principle of many of the practices of the scientific communication system. It applies for example to the procedures by which papers are accepted and refereed for formal publication (§4.5), as well as to the conventions of discussion and controversy during 'informal' communication at scientific meetings. It is also invoked as the principle behind the meritocratic tradition in the award of preferment or prizes for scientific achievement (§5.5). The legitimacy of this norm is obviously emphasized by reference to blatant counter-examples, such as the damage done to science when 'Aryan' or 'Marxist' scientists were given preferential access to the channels of communication and authority.

But there is also an uneasy awareness by scientists that this norm is not really consistent with the social differentiation of the scientific community into distinct specialties (§5.3), and strata of authority (§5.5). There is a tendency for groups of specialists to discriminate against the opinions of outsiders and lay-persons, whilst the Matthew effect often gives too much scientific weight to the views of members of the scientific élite.

Disinterestedness: *Science is done for its own sake.* That is to say, scientists should undertake their research, and present their results with no other motive than the advancement of knowledge. They should have no personal stake in the acceptance or rejection of any particular scientific idea. This norm thus underlies the convention that academic scientists should not be paid directly for their primary contribution to knowledge – a convention that is not altogether consistent with the payment of consultancy fees to scientific experts, or with the terms of most contracts for commissioned research (chapter 14). Any scientist who tried to bolster up the case for a discovery claim with data that were deliberately misleading (even if not downright fraudulent) could be accused of contravening this norm, which thus demands high standards of honesty amongst scientists.

Indeed, interpreted more symbolically, this norm forbids any open manifestation of the psychological commitment that scientists usually feel to their own discoveries. From this can be derived the impersonal objective style of scientific communications (§4.3), the muted tone of public controversy, and the self-effacing stance that scientists are expected to adopt when making discovery claims. This is why open disputes over priority are regarded as somewhat scandalous, since they uncover the interests that scientists have in the intense competition within the scientific community for personal recognition and rewards (§§5.1).

Originality: *Science is the discovery of the unknown.* That is to say, scientific research results should always be novel. An investigation that adds nothing new to what is already well known and understood makes no contribution to science. This norm lays emphasis on the discovery element in scientific epistemology (chapter 2). It enjoins upon scientists diverse forms of 'creative' behaviour and 'imaginative' thought. Originality is, of course, an obligatory condition for the publication of a research paper (§4.2), the acceptance of a PhD thesis, the award of a prize (§5.1), or almost any other act of recognition in academic science. Conversely, this norm strongly censures all forms of scientific plagiarism – i.e. passing off some other scientist's work as one's own – and forbids the communication of the same research result to several different primary journals at the same time. The norm of originality gives rise to the tendency to undervalue the technical and administrative services that are needed in the research process (§11.2), and to give little scientific weight to routine investigations associated with industrial production, technological development and other practical concerns.

Scepticism: *Scientists take nothing on trust.* That is to say, scientific knowledge, whether new or old, should be continually scrutinized for possible errors of fact, or inconsistencies of argument. Any justifiable critical comment should at once be made public. This norm institutionalizes a context of validation (chapter 3) within the scientific community, enjoining strict intellectual discipline and high critical standards on all scientists. This is evident in the peer review of communications (§4.5) and of research grant applications (§14.4), in the tradition of informal debate at scientific meetings (§4.7) and in all the other procedures by which discovery claims are accredited (§4.6). Scientists express their chagrin when this norm does not seem to have been conscientiously observed, as when a grave fallacy has passed unnoticed for a long time, or when dogmatic education seems to have blinkered them to important new discoveries.

This list is not quite as Merton originally presented it. Although scepticism is strongly encouraged within the scientific community, it is not organized very systematically, and is given very little direct recognition as a distinct personal trait. For this reason, Merton's definition of this norm as *organized* scepticism is slightly misleading. On the other hand, he did not, at first, include *originality* as a primary norm, even though this is an essential characteristic of all academic science. Originality and scepticism are, in a sense, polar opposites, which are difficult to satisfy simultaneously. They generate between them a dialectic – the creative/critical tension – which is the psychological and sociological counterpart of the hypothetico-deductive tension of the traditional philosophy of science (§3.7). As we shall see in the next chapter, this provides the cognitive dynamics by which scientific knowledge is forced to change.

The revised list of Mertonian norms – Communalism, Universalism, Disinterestedness, Originality, Scepticism – forms an appropriate acronym – *CUDOS*. We are thus reminded of the academic slang word, *kudos*, meaning 'glory, fame, renown', which must surely be the true reward for those who would sincerely obey these norms in practice.

6.3 An ethos of academic science

The Mertonian norms are much debated by sociologists of science. One school of thought treats them as essentially well founded in principle, and to a considerable extent verified by observation: other schools dismiss them altogether, on theoretical and/or empirical grounds. This is obviously an important debate, since it concerns the theoretical framework for a large part of the subject matter of this book.

The Mertonian scheme can be accepted simply as an empirical generalization covering the observed behaviour of academic scientists (cf. §2.3). It is obviously not implausible as a very broad description of the general pattern of rules governing the scientific way of life, at least in an idealized way. But we can also point to some very significant cases where particular norms are systematically flouted. For example, the norm of communalism is ignored on a very large scale by all scientists who participate in commercial and military research, where secrecy is the rule, rather than the exception (§12.5). Similarly, it seems almost laughable to suggest that research workers are 'disinterested' in the reception of their results, about which they obviously care so deeply.

On the other hand, it is instructive to imagine what science would be like if any of these norms were entirely inoperative. If that debatable norm of disinterestedness were further relaxed, the scientific communication system might be opened to straightforward advertising: how much could be believed if the research literature were full of slogans like 'Einstein's Theory is Best: Enjoy New Creative Powers with the Aid of this Latest Produce of the Greatest Mind in the Business'? The tragic experience of Soviet genetics under T. D. Lysenko is direct evidence of what can happen when the norm of universalism is not respected. Again, suppose that scientists were not expected to produce original results, but were permitted to spend their time ritually performing old experiments, or solving standard examination questions whose answers were already known: this would obviously be a travesty of research activity, and quite unrecognizable as science in the usual sense. Such arguments by *reductio ad absurdum* demonstrate the necessity of postulating something like the Mertonian scheme if we are to distinguish between scientific behaviour and other typical forms of social action.

But a purely descriptive account would miss the essential point. Merton proposed this particular set of norms as the main constituents of a coherent *ethos* of science. They define an *ideal* pattern of behaviour, which scientists should endeavour to follow. This ethos is more or less consistent in itself, but it inevitably conflicts with a variety of other personal and social considerations, and can seldom, therefore, be practised in full. The first three norms, for example, demand considerable sacrifices of personal interests in favour of the common good. In the short term, it surely pays to exploit a discovery in secret, or to exert undue influence as an 'authority', or to offer one's opinions for hire. Cases of non-observance do not necessarily disconfirm ethical norms: an ethos that did not, by implication, define its characteristic 'temptations' would be quite vacuous. The attempt by some sociologists of science to codify some 'counternorms' of scientific behaviour does not, therefore, invalidate the Mertonian scheme.

This takes us, of course, to another highly questionable standpoint in social theory, raising immense issues of moral philosophy and social psychology. But it also opens up a number of fruitful empirical and theoretical questions about how academic science works as a social institution. These are questions that it is easier to ask than to answer, but they are well worth asking all the same.

6.4 Does academic science have an ideology?

The first obvious question is whether this particular set of norms is 'necessary and sufficient' as an ethos for science. Perhaps some further normative principles should be formulated to describe or explain certain aspects of scientific behaviour that are not covered by the CUDOS scheme. For example, 'curiosity' (§2.5) is rightly emphasized as a scientific virtue, closely linked with the importance of being able to do research primarily 'for its own sake'. It has been suggested, therefore, that scientists should hold to a norm of *autonomy*: they should behave so as to give each other the utmost freedom in setting the objectives and methods of their individual investigations. Needless to say, this norm which would legitimate the extreme individualism and competitiveness of the traditional scientific life (§5.2), stands in opposition to many authoritarian and collectivist tendencies in modern science.

It is natural to enquire how this ethos is imposed on scientists, and how well it is enforced. What are the rewards for observing the norms, and the penalties for not observing them? The Mertonian scheme is so generalized that these questions cannot be answered in the abstract. In practice, it comes down to asking what happens if a scientist breaks a particular rule of scientific work, such as knowingly publishing false data. But these penalties often seem so weak, and are imposed so inconsistently from case to case and from rule to rule, that it is not clear that they can be related to any general motive except the desire not to lose communal esteem. Perhaps the system works more by custom and convention than by realistic calculation of the personal costs of conforming. Scientists are undoubtedly able to learn to follow the rules in detail, without being conscious of the general principles from which these rules are derived or by which they are enforced.

The reality of this ethos as a generalized account of scientific behaviour is also called into question when one enquires whether it is genuinely internalized by each individual, or whether it is no more than a codification of outward social forms. It has often been asserted that scientists have by nature, or acquire by nurture, a 'scientific attitude', which makes them peculiarly honest, objective, independent of mind, sceptical, rational, etc. (§15.1). The academic ethos certainly encourages scientists to exhibit such virtues, but this may be merely the posture they adopt when they are playing out public scientific roles, such as communicating their results, addressing conferences, or awarding each other prizes. There is much evidence to suggest that this apparent conscientiousness does not transfer automatically to other social activities and relationships and that scientists do not show this 'scientific attitude' in the relative privacy of the laboratory, the study, or the committee room.

We have already noted the connection between the norms of originality and scepticism, and the hypothetico-deductive methodology whose practice they encourage (§3.7). It has even been suggested that the Mertonian scheme should be extended to include *technical* norms, obliging scientists to replicate experimental data, attempt to falsify hypotheses, etc. Our study of the research process shows that this would be difficult to justify epistemologically, but there is obviously a close connection between the social ethos of science and the metaphysical assumptions of scientific work (§3.10). In order to behave satisfactorily as a scientist, one is practically bound to believe in the existence of the external world, the possibility of discovering some order in nature, etc. If we insist that the scientific epistemology has both

6.4 Does academic science have an ideology?

personal and communal components, we must bring together the cognitive and social regulative principles of the research process. Can these principles all be derived from a single general characterization of science, such as the 'rational consensus' principle (§1.5)? Despite the intellectual immodesty of such a grand question, it helps to show up the inseparability of the philosophical and sociological components of science studies (§8.6).

It also raises the very old question of whether there is some simple demarcation criterion between science and other bodies of organized knowledge. Do the Mertonian norms, for example, apply only to natural scientists, or do they constitute a more general ethos covering all the scholarly activities of academia? It is now customary for academic technologists, social scientists and humanists to order their communal affairs along essentially the same lines as physicists and chemists, even though the regulative principles governing evaluative scholarship, such as literary criticism, are not at all the same as those of empirical and theoretical research (§16.4).

This *academic ethos* is not unrelated to some of the norms of society at large. There is a remarkable consistency between the Mertonian norms and the 'human rights' adumbrated by political philosophers and codified in various international conventions. Thus, for example, 'communalism' is protected by the right of free speech, and 'universalism' is implicit in political and legal prohibitions on religious and racial discrimination. Merton's original paper emphasized the intimate connection between a 'free' society and a 'free' science, whilst other writers such as Michael Polanyi have presented the 'republic of science' (§6.1) as an enlightened model for society at large. A political philosopher might ask whether the scientific ethos had not been enlarged into an *ideology*, invoked to justify a wide range of social and intellectual activities.

But this harmony between the internal norms of science and some external social norms could also be seen as a simple consequence of history. The academic ethos defines the type of science that would be permitted in an open pluralist society: in another type of society, with a different political ideology, there would probably have arisen a different type of science, with a different ethos (§8.3) and different notions of 'truth'. This raises a fundamental issue, to which we shall eventually return: is the scientific community, bound together by its academic ethos and metaphysical assumptions, a compact, stable and unique social institution? In later chapters, we shall be observing the tendency of external social forces to influence and transform the internal sociology of science towards much more bureaucratic and hierarchical models (§12.5). The academic ideology might then be seen as a rationale for the relatively privileged social position of a scientific élite, legitimating their autonomy as 'honest seekers after truth', justifying the rewards for their expertise and creativity, and minimizing their personal responsibility for any ill effects of doing research 'for its own sake'. Without endorsing this cynical critique of science as an entrenched 'Establishment' (§5.6), we can agree that any serious analysis of the way in which people behave as scientists cannot be separated from our understanding of the way in which society is structured and functions as a whole.

Further reading for chapter 6

The original papers on norms are reprinted in
> R. K. Merton, *The Sociology of Science*. Chicago: University of Chicago Press, 1973 (pp. 256–78 and 286–324)

Merton's scheme is put in a broader sociological context by
> N. W. Storer, *The Social System of Science*. New York: Holt, Rinehart and Winston, 1966

but is criticized theoretically and empirically by
> L. Sklair, *Organized Knowledge*. London: Hart-Davis, MacGibbon, 1973 (pp. 102–82)

and by
> M. Mulkay, *Science and the Sociology of Knowledge*. London: George Allen & Unwin, 1979

Empirical evidence of deviations from ideal scientific behaviour is presented by
> M. Mahoney, *Scientist as Subject: The Psychological Imperative*. Cambridge, Mass: Ballinger, 1976

For an account of some of the more spectacular deviations from the scientific ethos, see
> W. Broad & N. Wade, *Betrayers of the Truth: Fraud and Deceit in the Halls of Science*. London: Century, 1983

This book is unscholarly and highly sensationalized, but contains valuable sociological insight on some aspects of scientific behaviour.

The norms of academic science are related to the consensus principle by
> J. M. Ziman, *Public Knowledge*. Cambridge: Cambridge University Press, 1967 (pp. 77–101)

7

Change

'Scientific revolutions are not *made* by scientists. They are *declared* post factum, often by philosophers and historians of science rather than by scientists themselves.'

Hendrik Casimir

7.1 Cognitive change

The world's scientific archives acquire something like a million new scientific papers a year. The growth in the quantity of scientific information was, until quite recently, an accelerating process. The number of papers published annually has been increasing exponentially for the best part of 300 years. An elementary calculation shows that this corresponds to an annual growth rate of about 5%: that is to say, the amount of new scientific information reported each year has been doubling about every 15 years, since the late seventeenth century.

The steady growth in its contents and in the scale of its operations has very important implications for the place of science in society (chapter 11). It is also one of its major internal characteristics. At this rate of growth, for example, half the information in a scientific archive must be less than 15 years old. Perhaps only a tiny fraction of this information is scientifically interesting or novel. Much of it will consist of factual data on very minor topics, recorded at higher levels of precision than previously. Nevertheless, unless the norm of originality (§6.2) is being systematically violated, scientific knowledge is changing rapidly by the sheer accumulation of new information.

This change is not simply additive. Current scientific instruments, techniques and theoretical concepts evidently differ profoundly from those of, say, 50 years ago. These intellectual resources are not all being applied to the solution of quite new problems which have never previously been investigated. In accordance with the norm of scepticism (§6.2), discovery claims and theoretical explanations are always subject to criticism and modification. It is of the very essence of the hypothetico-deductive method (§3.7) that knowledge that was once acceptable can be disconfirmed, made less credible, and hence, eventually superseded.

A science is thus unlike almost any other body of organized knowledge, such as a religious creed, in that its effective contents may be transformed almost out of recognition in the course of a few decades. In the opinion of most scientists, this transformation, whether by the accumulation of new data or by the supersession of old theories, must always be for the good. Current scientific knowledge is always considered superior to past knowledge, so that science must always seem to be *progressive*. This opinion is so firmly incorporated in the regulative and normative principles of scientific work (§3.10, §6.3) that it can scarcely be analysed as a distinct proposition: by definition, no research report or discovery claim in any field of science would come to be accepted as 'established knowledge' (§3.8) unless it were thought to mark an advance, or contribute to 'progress' in that field.

The nature of scientific change is, of course, the central topic in the study of science as an historical process. In the traditional histories of the sciences, the advancement of knowledge is largely attributed to the discoveries and insights of particular individuals, each building upon the work of his or her predecessors. This account is entirely consistent with an individualist ethos of academic science (§6.4) and with the notion of a unique external world that can be explored and mapped objectively by their separate efforts (§3.10). The sources of scientific change are sought in the psychology of personal creativity (§15.1), expressed in each instance within the context of the scientific knowledge available to the particular researcher.

Scientific biography is one of the great scholarly arts. There is much to be learnt about science from a learned and sympathetic account of how some great scientist, such as Michael Faraday or Louis Pasteur, came to have such a profound effect on the science of his time. But its main lesson is the variety of particular circumstances that can influence the outcome of any scientific investigation. As in every other branch of history, general considerations are difficult to detect amongst the diversity of personal factors in the lives of the individual actors.

In practice the discovery process is seldom recorded in such fine detail that all these factors can be satisfactorily assessed. The personal psychological factors, for example, are seldom better than conjectural. The tendency in the history of science is to concentrate on the *cognitive* context within which research was undertaken and discoveries made. For want of better information, the historian outlines this context in terms of the books that the researcher is said to have read, or the informal communications (§4.7), such as letters from other scientists, that happen to have been preserved. This sort of information is obviously highly relevant to any explanation of scientific change. But it defines the situation facing each scientist as an intellectual 'problem', for which the discovery was then an attempted solution. That is to say, the historical event is projected on to the plane of philosophy, without regard to its psychological or sociological dimensions.

Philosophical analysis of the historical development of scientific knowledge is very instructive. The various philosophies of science give rise to various models of scientific change which can be compared with the facts of history. If, for example, the basic 'method' of science is strictly hypothetico-deductive (§3.7), then numerous episodes of 'conjecture and refutation' should have occurred. If, on the other hand, the progress of science is best represented by an evolutionary model, there should be evidence of many 'mutant' theories which looked viable enough at the time but which simply failed to survive in the competition for accreditation (§4.6). In this sort of discussion, the actual conceptual resources available to a particular discoverer have to be analysed very carefully, so that great attention is paid to the scientific theories of the day, their logical coherence and their consistency with known facts. But we can never imagine ourselves back into a past state of ignorance and error. As a consequence, it is almost impossible to write an 'internalist' history of science, that does not exaggerate the rationality of every successful step that was taken. In the light of hindsight, the way forward seems much too clear and simple – and most of our predecessors seem absolute nitwits not to have seen it.

7.2 Institutional change

Philosophical models of scientific progress not only underplay the erratic, irrational, personal element in the process of discovery: they also ignore sociological factors, such as the influence of the general cultural milieu and the demands of technology (chapter 9). In particular they do not take account of the fact that the contents of a body of information are not independent of the social organization of those who know this information and who may have an interest in preserving it or changing it. Even an academic science such as astronomy, with no applications in view, does not transform itself internally by its own intellectual dynamic, as if all that happens can be explained in terms of facts and theories, concepts and methodologies. Scientific change also involves university departments, professional appointments, learned societies, publishing houses, educational curricula, and other institutions where scientists have significant rôles.

It is obvious that major scientific discoveries give rise to institutional development and change. An exciting conceptual or technical breakthrough opens up a whole new range of scientific problems (§2.15). Thus, for example, the determination of the structure of DNA in the early 1950s suggested the possibility of solving the genetic code, and investigating in molecular detail the mechanisms of metabolism, growth and heredity in living cells. A whole new generation of scientists, trained in chemistry and physics, took up these problems, and soon became highly proficient in their investigation and solution. As this new 'problem area' expanded on the cognitive map, the new scientific specialty (§5.3) of molecular biology began to emerge as a distinct social entity. This specialty is now so large that it can be considered a new academic *discipline*.

The appearance and consolidation of new disciplines is one of the characteristic sociological phenomena of academic science. At first the emerging specialty is only observable as a nodal point in the network of citations (§4.2). Then scientists whose research is associated with this co-citation cluster organize little research conferences to discuss their common interests, or are commissioned to write articles for a special issue of a primary journal drawing attention to progress in this particular problem area. An 'invisible college' (§5.4) begins to condense out, in the form, say, of a semi-official association held together by further conferences, the regular exchange of pre-prints and re-prints (§4.7) and the publication of an informal 'newsletter'. In due course, the association develops into a regular learned society, whose newsletter has become a reputable primary journal. A hierarchy of authority (§5.5) is soon set up to preside over conferences, edit journals, allocate resources, and confer recognition on the members of the new discipline.

The final stage of institutionalization is the incorporation of this discipline into the educational curriculum. Teachers and students specializing in the new subject set themselves apart from their colleagues in neighbouring disciplines by insisting that they are following a distinctive intellectual tradition (e.g. 'We *molecular biologists* are not just *biochemists*') which can only be taught by individuals with the appropriate training, holding appropriate academic posts (e.g. 'Professor of Biophysics'). If the new discipline becomes the basis of a new practical profession (e.g. biotechnology), a professional organization may have to be established to regulate the qualifications of certified experts (cf. the Institution of Electrical Engineers).

The actual history of the institutionalization of a new scientific discipline is usually far more complicated than this schematic account. The main point is, however, that this process is going on simultaneously with the cognitive development of the new subject, and is not simply a consequence of that development. The advancement of science depends on the social advancement of scientists, and knowledge becomes 'established' (§3.8) by its connection with established scientific institutions (§5.6). If we are to give a satisfactory account of scientific change, we must allow for the reflexivity of the relationship between the cognitive and social dimensions of scientific activity.

7.3 Change by revolution

The close linkage between cognitive and social factors is evident in the general theory of scientific change put forward by Thomas Kuhn in 1963. This theory has had a very wide influence in science studies. Drawing evidence from various well-known episodes in the history of science, Kuhn argued that scientific change could best be

7.3 Change by revolution

described empirically in terms of relatively long periods of *normal* science sharply punctuated by profound *revolutions*. This historical phenomenon he explained as follows.

Normally, a field of science is dominated by a *paradigm*, which is accepted, practically without question, by the active scientists in that field. The central element in such a paradigm is a well-established and apparently comprehensive theoretical system – for example, Newton's laws of mechanics and gravitation, which ruled over classical physics for 200 years. Associated with this theory, one finds paradigmatic research methodologies and problem-solving techniques: the Newtonian methodology, for example, is the study of the motion of macroscopic bodies, as in astronomy, using the mathematical techniques of the differential calculus. The various elements of the paradigm are closely articulated, and are taught to students as a firmly established framework for research. As a consequence, most research is directed towards the investigation of 'puzzles' which can be defined and solved within that framework. Thus, the Newtonian paradigm suggested numerous puzzles concerning the details of the motion of the Sun and the planets, under the influence of their mutual gravitational attractions.

As the research proceeds, however, various *anomalies* become apparent. Certain observable phenomena seem to be inexplicable within the theoretical and technical framework of the paradigm. In the case of classical physics, the notorious anomaly was the rotation of the perihelion of the orbit of the planet Mercury, which seemed much larger than the calculated gravitational effects due to the other planets. Despite every effort, these anomalies resist conventional analysis. It then begins to dawn on people that the current paradigm is not adequate to deal with them, and will have to be drastically modified or even abandoned altogether. Research thus moves into a *revolutionary* phase, where very speculative hypotheses are proposed (§2.14) in the hope of resolving these difficulties. This flurry of activity soon leads to the discovery of a comprehensive new theory that fits the previously accepted facts and is also successful in explaining the anomalies. After a suitable process of criticism and corroboration (§3.6), there is general agreement that this new theory must be essentially correct and it becomes the basis for all further research in the subject. In other words, a new cycle of normal science has begun, with the new theory as the central element of a new paradigm. In our historical example, Einstein's General Theory of Relativity has displaced Newtonian mechanics as a comprehensive account of dynamical phenomena on the astronomical or cosmological scale, and is being used to solve a whole range of new scientific 'puzzles' about black holes, the expansion of the Universe, gravity waves, etc., which simply could not have been imagined within the Newtonian framework of thought.

This is a grossly oversimplified account of a model of scientific change that was originally presented in much more judicious terms, and has since been considerably modified and elaborated by the author and his various critics. Some of this critical literature fails to do justice to the issues at stake. Many students of metascience are given the impression that it is their duty to decide whether they are 'for' or 'against' Kuhn – in exact proportion, it seems, to their being 'against' or 'for' Popper! In view of the fact that the Popperian policy of 'conjecture and refutations' is precisely the Kuhnian process of 'revolutionary science', this polarization is difficult to understand. The notion of a 'paradigm' has proved difficult to pin down, and some philosophers of science would follow Imre Lakatos in referring, rather, to a *research programme*, with an inviolable core of basic principles surrounded by more disputable beliefs that are open to corroboration or disconfirmation. From a philosophical point of view, this distinction may well be justified, but it makes little difference to the discussion of the 'Kuhnian paradigm' in science studies, which will take up the remainder of this chapter and much of chapter 8.

7.4 *The historical structure of scientific revolutions*

Is Kuhn's theory of scientific change well founded? Is it in accord with the historical facts? The above account of a typical scientific revolution is very schematic, but does it represent roughly what happened in a number of instances in various fields of science at various times in the past? As with many highly simplified models of social phenomena, the supporting evidence is plausible enough, but is gravely compromised by contrary considerations.

There is nothing new, of course, in the notion of a 'revolution in thought': the quasi-political metaphor goes back to at least the seventeenth century, and has been applied to innumerable episodes in the history of science. As in many political revolutions, the real question is whether there has, in fact, been a discontinuous change of régime, where an old theory has been swept away and entirely replaced by a new one. This has, indeed, sometimes happened, as in the case of the phlogiston theory in chemistry, and the caloric theory of heat.

Nevertheless, it is easy to exaggerate the authority of a new intellectual scheme, and to ignore the good scientific work that still goes on under the traditional paradigm. Relativity theory, for example, undoubtedly 'revolutionized' mechanics and electromagnetism, but it really had almost no effect on some of the major fields of research in classical physics, such as hydrodynamics, where the Newtonian paradigm still rules. Intellectual purists can argue that this apparent continuity is misleading, because all the concepts of classical physics are now explained differently in relativistic language (§2.13) and have thus become 'incommensurable' with their previous meanings. This is a valid philosophical point, but it could be merely a semantic change reminiscent of the political practice of changing the names of the streets of a city after a successful coup.

It must be admitted, moreover, that science does not always develop by a

7.4 The historical structure of scientific revolutions

succession of revolutions. There have been other patterns of change, represented by other quasi-political metaphors. After the 'breakthrough' of the determination of the structure of DNA, the virgin territory of molecular biology was rapidly *colonized*. In this process, microbiology and virology were *annexed* by physicists and chemists. In the late nineteenth century, physical chemistry *seceded* from chemistry to form an independent discipline, but nowadays a *merger* is taking place between the traditional preclinical subjects of anatomy and physiology. In some fields of science, an economic metaphor is more appropriate as in the remarkably steady *economic growth* of accelerator design in high-energy physics. The history of science should not be sliced up into revolutionary episodes just to fit Kuhn's model.

The theory quite properly takes account of the dogmatism of science education (§16.2), which often seems designed merely to reproduce the consensual status of the 'established' knowledge of its day (§3.8). There is also abundant historical evidence for an almost pathological psychological resistance by many scientists against the 'paradigm shift' needed to see their subject in a new light – witness the failure of Wegener to persuade the geologists to take seriously his Theory of Continental Drift. These features of the model are so familiar from ordinary academic experience that they scarcely call for historical validation.

Nevertheless, the predominance of this indoctrination and intransigence may be exaggerated. Until the last century, there was very little direct education in science, anyway. The modern researcher graduates through elementary, secondary and tertiary science education into the more advanced milieu of the graduate school or research institute, where there is much more controversy and uncertainty in the intellectual atmosphere. Experience in research teaches scepticism towards what was supposedly well established, as well as towards novel discovery claims or theoretical speculations. It is not only the up and coming younger people who are the most open to a new point of view. The reception of Darwin's theory of evolution, for example, was very mixed; although it roused fierce opposition, it also quickly won the support of a considerable number of highly reputable members of the scientific 'establishment' of the day. Thus, the psychodynamics of Kuhn's model may not be quite as straightforward as he suggested.

For many philosophers and scientists the most controversial feature of the theory is the implication that science 'normally' consists of solving puzzles by routine methods. Here the historical evidence supports Kuhn strongly, in that research usually proceeds by the formulation of a succession of problems (§2.15) which must be sufficiently well posed to be solved by the techniques available. Almost every achievement of every scientist has been built upon, and embedded in, the achievements of other scientists (§4.2); there is no escaping the humbling truth that we have only been able to 'see a little further' because we 'stand on the shoulders of giants', as Newton put it, echoing an ancient maxim.

Nevertheless, Kuhn's characterization of 'normal' science is easily misunderstood.

He is not suggesting, of course, that this is equivalent to the pedagogic practice of doing lots of contrived exercises or artificial puzzles in order to gain technical skill. The very essence of a research problem is that the answer is not known to anyone. A research result that is not 'original' is simply not publishable (§4.2). The mainstream of western science has never strayed very far from its basic norm of originality in that limited sense. The real question is whether scientists are 'normally' too timid in the investigations that they undertake and only set themselves research problems with very limited objectives. Does the radical spirit go out of science, except during occasional revolutionary periods?

Here is where Kuhn's notion of a paradigm as a complete system of detailed thought calls for analysis. Even a single over-arching theoretical system, such as Newtonian mechanics, does not provide all the specific research methodologies and problem-solving techniques that are actually needed to answer all the particular scientific questions that lie within its scope. Numerous sub-theories, sub-methodologies, and sub-techniques have to be developed (by the usual methods of science!) to cover all the specialties and sub-specialties (§5.3) into which the discipline as a whole has become differentiated. That is to say, each specialty develops its own 'sub-paradigms', within which 'normal' research can supposedly proceed. In practice, however, these sub-paradigms are often inconsistent with one another, or only weakly validated. Thus, what may look from the outside to be a very timid, routine piece of research may be intended to test severely an 'accepted' fact or theory, may turn up an 'anomaly', and may give rise to a minor 'revolution' in that specialty.

In fact, the philosophical machinery of the model is too tidy. The notion of an 'anomaly' is familiar enough in research experience, but it is not the only spur to speculative thought or radical experiment. The search for some link between disjoint theoretical domains has been a powerful agent of scientific change – witness Einstein's General Theory of Relativity, which was not originally motivated by a desire to explain the 'anomaly' in the motion of the perihelion of Mercury. In other cases, the revolutionary hypothesis has already been put forward, before the 'anomalies' are discovered that seem to make it necessary: this could be said of Wegener's Theory of Continental Drift, which was eventually forced upon geology by the discovery of 'anomalies' in the magnetism of the rocks. Like the Popperian model of successive conjectures and refutations (§3.7), the Kuhnian model of paradigms and anomalies is not unfaithful to some features of the research process, but does not cover all the considerations that lead scientists to undertake particular investigations or to accept as valid particular scientific results.

To sum up: a detailed study of the history of science will always reveal a tense dialectic between conservatism and radicalism. This dialectic is present in the breast of the individual scientist, who may honestly say, 'On Monday, Wednesday and

Friday, I do "normal" science: on Tuesday, Thursday and Saturday, I do "revolutionary" science', indicating that no serious piece of research is totally routine or totally novel. In any field of science, one may, from time to time, observe a punctuated evolutionary sequence of 'normal' and 'revolutionary' phases, as one or other of these tendencies takes the upper hand. This phenomenon can occur on any scale, up to the very largest, where a whole scientific discipline may undergo a revolutionary transformation. Nevertheless, this is only one of the many ways in which scientific knowledge grows and changes.

7.5 The sociodynamics of scientific life

Kuhn's theory of scientific revolutions does not give an adequate account of the history of science in all its diversity. Nevertheless, it has contributed immensely to the internal sociology of academic science by drawing attention to a number of sociodynamic effects which had previously been largely neglected. An 'invisible college', for example, must be seen as something more than a 'coterie', or 'clique' of scientists who happen to come into contact through their membership of a particular intellectual specialty (§5.4) and who then get together more formally to establish communication links, allocate resources, establish a stratification of authority (§5.6), and so on. Through their shared interest in a particular aspect of nature, they quickly develop a common intellectual tradition, they acquire a preference for particular techniques of investigation, and come to accept – even if only provisionally – certain empirical facts and conceptual schemes concerning their subject. In other words, there is created what Ludwik Fleck has called a 'thought collective', whose members share a characteristic 'style of thought'.

By shifting the emphasis thus from the 'paradigm' as an intellectual abstraction to the 'thought collective' as an identifiable social group, we uncover the source of its influence on the individual. Kuhn draws particular attention to the effects of technical training, but within such a group there are many other social pressures towards conformity. Not surprisingly, its members quickly learn to see things the way other members do, and thus internalize conceptual frameworks that they find it very difficult to escape from, even when new facts point to the need for new points of view.

This conformity to the collective 'style of thought' is particularly evident in the formulation of research plans. Those who share a research *tradition* also share a characteristic set of research *problems*, whose solution would thus constitute a shared research *programme*. This programme may not be publicly articulated; the social pressure for conformity may be very slight; nevertheless, the imagination of the individual researcher may be extraordinarily hampered by this scarcely conscious influence. The phenomenon of simultaneous discovery is evidence of the fact that it is in their choice of research problems that most scientists show themselves to be most under the thumb of their specialist paradigms.

But the task of a thought collective is to carry out its research programme, which means, in the end, that it must work itself out of a job. If all that it succeeds in doing is to validate a pre-existing paradigm, then its members begin to realize that only 'puzzle-solving' is left, and many move to new fields where there is more opportunity to win recognition by showing scientific originality (§5.1). If, on the other hand, there is scientific progress – a breakthrough, a revolutionary change – the group is likely to break up, as the specialty expands to take in new members, or develops a new cognitive structure.

If a thought collective were merely an informal social group, this restructuring could be accommodated without severe strain. But if the invisible college has existed for some time, it will almost certainly have been institutionalized, through academic appointments, educational curricula and the foundation of research institutes. Many of its members will have acquired substantial research resources and positions of authority (§7.2). For reasons of social expediency, persons with a large stake in the collective will strongly resist cognitive change. Scientists are all too human. We may not be very sympathetic to the scientist who says 'If this new theory is accepted, then I shall lose my standing as an authority, and all my accumulated know-how – therefore I refuse to believe it, and will argue against it'. But from a sociological point of view, this is merely the contrary rôle to that of the scientist who makes an ill-justified claim to a discovery, or launches a highly speculative hypothesis, in the hope of winning fame and fortune in the same public arena. Each is following the logic of a social situation.

Our understanding of the sociodynamics of scientific change is very incomplete. Perhaps this is because the ultimate purpose of the scientific enterprise is not just to do research but to produce scientific progress – that is cognitive change. This ideology of permanent revolution may be quite congenial to some people, but it is incompatible with the stabilising norms of ordinary social groups. Academic science, being a collective activity carried out by quasi-autonomous individuals, thus preserves an uneasy balance between conservative and radical tendencies, and changes in a very erratic manner under the influence of both cognitive and social forces.

Further reading for chapter 7

The books by Meadows, Crane, Price, and Menard recommended in connection with chapter 5, give background information on the growth of specialties as cognitive clusters and as social institutions.

The classic text, setting out the theme of this chapter at length, is

T. S. Kuhn, *The Structure of Scientific Revolutions*. Chicago: University of Chicago Press, 1962

7.5 The sociodynamics of scientific life

A brilliant and evocative work, first published in German in 1935, which already contained many of Kuhn's ideas – and much more besides – is now available in English:

L. Fleck, *Genesis and Development of a Scientific Fact*. Chicago: University of Chicago Press, 1979

Some well-known essays stimulated by the controversy between Kuhn and Popper are published in

I. Lakatos, & A. Musgrave, (eds), *Criticism and the Growth of Knowledge*. Cambridge: Cambridge University Press, 1962

In particular, this volume contains (pp. 91–195) the paper by Lakatos outlining his theory of 'research programmes'.

An evolutionary account of scientific change is given by

S. Toulmin, *Human Understanding*. Vol. 1. Oxford: Clarendon Press, 1972 (pp. 1–129)

Scientific change is represented in terms of problems and 'research traditions' by

L. Laudan, *Progress and its Problems*. London: Routledge & Kegan Paul, 1977

An anarchistic swipe at all formal models of scientific method and scientific change is taken by

P. Feyerabend, *Against Method*. London: Verso, 1975

A succinct review of the subject, with wider references, is

G. Böhme, 'Models for the Development of Science'. In *Science, Technology and Society*, ed. I. Spiegel-Rösing and D. de Solla Price, 319–54. London: Sage, 1977

8
The sociology of scientific knowledge

'The real truth never fails ultimately to appear: and opposing parties, if wrong, are sooner convinced when replied to forbearingly than when overwhelmed.'

Michael Faraday

8.1 Science and the sociology of knowledge

The Fleck–Kuhn account of scientific change (§7.5) suggests a more radical approach to our whole subject. Instead of starting with a philosophical perspective (chapters 2 and 3), which emphasizes the cognitive aspects of science, we should perhaps have taken a sociological point of view from the beginning. In the past decade, academic metascience has been greatly influenced by a research programme which looks on science as *primarily* a social institution. This programme stems from the more general discipline of the *sociology of knowledge*, which used to be concerned mainly with the place of social-science knowledge in the culture of a particular type of society, but which is now being turned on the natural sciences and their associated technologies.

A programme of this kind is clearly implicit in what has already been said in previous chapters. The historical course of development in any field of science has a significant social component. The rate of scientific change, for example, is strongly influenced by the disciplinary structure of the scientific community, and not simply by the scientific ideas that happen to be current. As we shall see in later chapters, external societal forces such as technological needs also affect the direction of scientific development. Indeed we are forced to go considerably beyond Kuhn's model of scientific revolutions (§7.3), which fastens on the cognitive paradigm as the major social factor in scientific change. Any particular scientific development is *indexical*: it is indelibly labelled by the intellectual, technical, and political context in which it arose. Even a scientific genius cannot escape from the contingencies of time and place. However novel a scientific hypothesis may be, it is necessarily limited by the technical capabilities of its epoch, and can only be fashioned out of themata (§2.14) that are already familiar.

Historians of science have never denied this truism in principle, although they have often blinkered themselves by looking only for internal intellectual factors in the advancement of knowledge. For this reason, it is valuable to extend the potential influences to a much wider range, including many general cultural and social factors that would not be recorded at all in the scientific archives. To mention a famous example: Charles Darwin's theory of the evolution of species by natural selection owed as much to his acquaintanceship with the practical craft of breeding domestic animals, and to his having read the economic treatise of Thomas Malthus, as it did to the research results of other naturalists, past or contemporary. There can be little doubt that the Darwinian revolution in biology was decisively influenced and shaped by the social milieu of Victorian England, with its characteristic agricultural and industrial capabilities, its characteristic class relationships, and its characteristic political and religious ideologies.

Nevertheless, even those historians and sociologists who have seen the intellectual culture of a particular society as an ideological reflection of its material culture have usually hesitated to describe its scientific views in the same way. They have usually accepted the traditional epistemology of science, and assumed that the logic of justification (chapter 3) was beyond criticism. They would recognize that a variety of factors are at work in determining the direction in which research is undertaken. If some of these factors were not at work, then certain discoveries might not be made or might appear in a very different guise. But in the end they would expect the result to be the same. If, say, Nelson had lost at Trafalgar, nineteenth-century Britain might not have combined industrial capitalism with maritime imperialism, and the circumstances that led both Charles Darwin and Alfred Russel Wallace to much the same conclusions might not have occurred. Nevertheless (according to the traditional view) a theory of biological evolution by natural selection would eventually have been proposed by somebody, and found scientifically acceptable, because the criteria for scientific validity transcend the particular circumstances of any historical epoch or social milieu. On this view, the determining factor would be how well the process of validation had been carried out. If this process had followed the recognized scientific methodology (§3.7), then false theories would eventually be refuted and sound conjectures would be corroborated. Whatever may happen in the short run, therefore, cognitive change in science is progressive (§7.1) in that successive theories must converge, in the long run, on an unchallengeable body of knowledge (§3.8).

8.2 Epistemological relativism

Unfortunately, philosophy has not lived up to its promise of providing a complete method for validating research claims. The standard procedures for eliminating scientific error still leave a great deal of room for intellectual manoeuvre. Even if we accept certain facts as well founded, we cannot use the principle of induction (§3.4) to construct a unique general law to cover them (§2.9), nor a unique theory to explain them (§2.10).

Scientific knowledge is essentially *underdetermined*: in principle, there are any number of possible interpretations of a finite set of observations. For this reason, it is impossible to demonstrate that all scientific knowledge must eventually converge on a coherent body of 'objective truth' about the natural world. This is a preconceived notion, not a necessary consequence, of the way that science is done.

Scientists are thus free to choose from the set of scientifically tenable theories the one that accords best with their non-scientific preoccupations. Such preoccupations may be idiosyncratic and unself-conscious, such as a taste for geometric rather than algebraic representations in theoretical physics. But they are most likely to derive from the social context in which the research is carried out – for example, quantitative data and commercial metaphors would naturally be used in a capitalist society as distinct from, say, religious metaphors in a less secular period. Thus, any body of scientific knowledge may contain a significant component that is socially determined, and hence *relative* to the particular social group that has created this knowledge.

Epistemological relativism would not be very disturbing if it were only a matter of allowing for local fashions in the choice of topics for investigation or of models to explain the results. But we are now facing a more serious issue than, say, the difference between the Newtonian and Cartesian research traditions in eighteenth-century England and France. According to some philosophers and sociologists, the whole status of Western science is at stake.

The trouble is that science does not even have an absolutely sure method for eliminating 'error'. The philosophers have not succeeded in defining a universal set of criteria for the validation of scientific theories. There does not seem to be a completely watertight procedure for denying 'facts' or for refuting theories. Tests of inconsistency that are regarded as compelling in one social setting are not always thought to be convincing in another country, or in another epoch. What counts as overwhelming proof of the cause of an illness amongst the Azande is treated as superstitious 'witchdoctory' by European doctors – and *vice versa*.

This incommensurability of standards of proof cannot be overcome by an appeal to some abstract notion of 'rationality'. It derives from the differences in the total world views of members of different societies. In effect, Kuhn's concept of a paradigm applies with peculiar force to the regulative principles of scientific work (§3.10). Western scientists appeal to these principles, which they regard as 'self-evident', whenever the more detailed practices of the scientific method come into question. But these principles cannot be 'proved', and there are perfectly sane people, brought up in other cultural traditions, who do not accept their force.

Most scientists, naturally enough, find this very hard to take. They realize that their research is fallible, and almost certain to be superseded, but would not admit that it might seem wrong-headed in principle. They cannot believe that a time might come when science might be done according to some different metascientific paradigm, when their own hard-won discoveries might be held up for scorn as 'mere superstition'. But radical epistemological relativism seems fundamentally unassailable by formal philosophical argument. Fortunately, like other forms of complete scepticism, it is much more disconcerting in theory than in practice. It does not really go much further than Hume's critique of induction (§3.4) in challenging the validity of received scientific theories, and does not significantly affect the reliability of current scientific knowledge in its own context. Nevertheless, it is the ultimate and decisive weapon against naïve realism, and other forms of philosophical scientism (§3.9).

8.3 The 'strong programme' in the sociology of knowledge

The threat to science from radical relativism is serious – but not dangerous. Scientists who say that they are happy to follow the traditional 'method' can safely continue their work in any society where the appropriate regulative principles are universally accepted. But what if scientists did not really practise what they preach? Some sociologists and psychologists have presented evidence that scientists simply do not observe the norms that are supposed to regulate their communal behaviour (§6.3). Since the methodology of academic science depends upon the systematic observance of these norms (§6.4), this would suggest that research claims are not being properly tested and validated. In other words, science should not be treated as if it were epistemologically superior to any other system of belief.

This is the basic principle of the so-called '*strong programme*' in the sociology of knowledge. This programme is not, in principle, antagonistic to science. It aims simply to study the conditions which bring about belief, without prejudice as to whether the belief is 'true' or 'false', 'rational' or 'irrational', or even 'successful' or a 'failure'. Thus, for example, considerable attention is given to highly controversial and marginal fields of science, such as research on gravity waves and even to *parascientific* enterprises such as the detection of extrasensory perception (§16.3).

There can be no objection to a programme of empirical research on such interesting and important matters. The notion of scrutinizing impartially *all* the conditions that give rise to a particular state of scientific knowledge is unexceptionable. Of course it is very bold to suggest that sociological analysis could account for the creation of the frailest and most intangible of social products – human knowledge – when the same sort of analysis has not yet succeeded in accounting for more robust and tangible social entities such as railroads and prisons. The strong programme is also, in a strict sense, self-contradictory, since it practically assumes the validity of the scientific approach to social 'facts' and 'explanations' which it is at pains to repudiate. But this reflexive inconsistency is only a formal objection, which need not stand in the way of research in this field. The strong programme at least has the virtue that it tries to pose well-defined questions, and to suggest answers which are capable of being tested by reference to ascertainable facts.

In particular, this sort of sociological study has clearly shown that the actual research process in the natural sciences is seldom carried out according to any of the formal canons of 'scientific rationality', thus fully confirming the 'fraudulence' of the scientific paper as an historical report (§4.3). It has also shown how difficult it is to pin down the 'scientificity' of conventional research in a field such as atmospheric physics, as distinct from a 'parascientific' activity such as the investigation of reports of unidentified flying objects. From this point of view, the privileged status of 'official' science would not appear to be earned by a more rigorous application of the 'scientific attitude' (§6.4) by individual researchers.

On the other hand, sociological investigations within this and similar research programmes have not demonstrated that what passes for well-founded scientific knowledge does not satisfy the basic scientific criteria of self-consistency and factual accuracy. That is to say, scientists do not simply 'manufacture' knowledge to order, and 'negotiate' interpretative schemes as if they were commercial contracts. Nature is not as malleable as all that, and the academic community does not function along the lines of an oriental bazaar, where anything can be made to go. The weakness of the strong programme is that it encourages and legitimates investigations that start with a partial list of the factors at work in science, and therefore arrive at very doubtful conclusions. There is a temptation, at times, to undervalue the strength of the scientific tradition, which is itself one of the pre-existing conditions that help to bring about scientific beliefs. In trying to avoid the obvious path of accepting ordinary 'scientific' arguments at their face value, the tough-minded sociologist may seriously overestimate the influence of social interests and other extrascientific considerations. There is no accounting for the origins of ideas without allowing for the creative and constraining powers of other ideas.

8.4 Science as a social enterprise

From the standpoint of the general sociology of knowledge, academic science is only one of the many sub-cultures of society. It is seen primarily as a social *institution*, connected more or less closely with other institutions, such as government or education, and subject to the usual societal conflicts of class and corporate interests. Science is thus supposed to differ from other sub-cultures only in its employment

8.4 Science as a social enterprise

of certain highly technical resources, and in its explicit rationality — although even the latter distinction is disputed by some sociologists and philosophers (§8.3).

The sociology of the scientific community, on the other hand, is concerned primarily with the distinctive internal structure of this institution. The scientific sub-culture is differentiated from other cognitive sub-cultures, such as the Law and the Church, by the unique set of rules and norms that academic scientists are expected to follow in the practice of their profession (chapter 6). The norms may be somewhat idealized, but the rules they sanction are not arbitrary: as we have seen (§6.4), if most scientists could not be relied on to abide by such rules, they would not be able to carry out research according to the 'method' sketched out in chapters 2 and 3. In other words, this tradition in the sociology of science is based upon the reasonable premise that a scientific community sharing the scientific ethos is an essential social milieu for satisfactory performance of the scientific rôle.

This perspective certainly discloses the intimate connections between psychological, philosophical and sociological factors in scientific activity. But it still gives primacy to the psychological and philosophical dimensions in the creation and validation of knowledge. Science is still seen as the work of individual scientists undertaking investigations, and solving problems in the pursuit of understanding and truth. The nature of science as a *collective* enterprise is not sufficiently emphasized.

The notion of science as 'public knowledge' (§1.5) characterizes academic science as *intrinsically* social. 'Communalism' becomes the primary norm of the scientific community (§6.2). Although the research process is fundamentally dependent on the tacit knowledge (§3.3) and individual skills that scientists bring to their work, this knowledge cannot properly be said to be 'scientific' until it has been made explicit and accessible to other researchers. However profound or well conceived a piece of research may be, however refined its concepts and techniques, it can only become a 'contribution' to science by being put into communicable form and submitted for publication (§4.3). An individual with scientific training may carry out research, make discoveries, formulate hypotheses and test them experimentally — yet these applications of the scientific 'method' would not be truly scientific work unless it were coordinated with the work of others through the formal and informal channels of scientific communication. Leonardo da Vinci made superb observations, but these never got published; Sherlock Holmes was immensely clever at solving problems, but his methods were too personal to be used by other investigators; Robinson Crusoe was extraordinarily ingenious, but until Friday came along he was in no position to communicate his technological discoveries to his fellow men: strictly speaking none of these intellectual paragons should be called a 'scientist'.

From this point of view, the communication system (chapter 4) is the essential structural component of the scientific community. This is what holds it together, and gives it meaning as a collectivity. Through this system, each scientist becomes an active participant in an enterprise whose objectives and achievements transcend the goals and capabilities of any single person. In the rhetoric of the academic ideology (§6.4), this transcendence is attached to *metaphysical* entities such as 'objectivity' or 'truth': from a sociological standpoint, such terms can best be seen as attributes of *collective* representations of life and nature. A scientist who claims to be 'an honest seeker after truth' is saying, in effect, that he or she is committed to public standards of credibility and irrefutability in reports of the results of research.

8.5 Establishing a consensus

The objective of academic science is to find out about the natural world. This is achieved by bringing together the research reports of individual scientists, and merging them in a collective account labelled 'established scientific knowledge' (§3.8). An essential characteristic of this account is that it is *public* – not necessarily in the sense that it is known and understood by all mankind, or by all adult citizens of a particular country, but that it is freely available in public documents, written in a public language and stored in public archives (§4.1). This account, moreover, is not sacrosanct: following the norms of originality and scepticism (§6.2), scientists are enjoined to communicate to the archive reports of further research of a corroborative or critical nature.

Scientific knowledge thus belongs essentially to what Karl Popper has called 'world 3' – the domain of general cultural resources which exist independently of any particular person. This fundamental feature of the sociological characterization of science is usually glossed over in the traditional philosophical analysis of the scientific 'method'. It explains, for example, the stress on *generality* in the scientific description of nature (§2.3) and in the theories that are produced to explain scientific laws (§2.9). A complete archive of all distinct facts and their specific interpretations would be utterly meaningless and useless as a collective resource for a community of any size.

The rule that scientific information should be communicated explicitly and unambiguously also has an important influence on the form and contents of scientific knowledge. It explains the significance of instrumentation (§2.6), quantitative observations (§2.7) and mathematical analysis (§2.13) in science, as distinct from sub-cultures such as literature where this rule does not apply. One might say, indeed, that scientists *must* be 'logical', because otherwise they might begin to make ambiguous or inconsistent statements, which could not be properly understood without recourse to private information about what was really meant, and would not therefore be acceptable as 'public' knowledge.

The social force of competition for recognition (chapter 5) drives academic science towards controversies which must eventually be resolved to establish a rational *consensus*. The 'method' that is used to resolve scientific controversies (§3.7) relies upon our intuitive feeling that we cannot tolerate any direct contradictions between different public accounts of nature, whether between one set of 'facts' and another, or between one theoretical model and another, or between 'facts' and 'theory'. If such a contradiction becomes evident, then new studies must be undertaken until everybody comes to agree on a single coherent account. This explains, for example, the rule that experimental results must be reproducible (§3.2), otherwise they would not be, so to speak, 'consensible'.

Does the rule of logical consistency derive directly from this sociological characterization of science, or does it just happen to be one of the regulative principles (§3.10) that all modern scientists actually bring to their work? It is hard for us to imagine that *scientific* knowledge could be self-contradictory – e.g. like Zen – but we might just possibly be subject to epistemological relativism (§8.2), and not realize the extent to which we are all under the sway of a paradigm of logicality. Perhaps this question should be put around the other way: what degree of logical inconsistency and lack of consensus is tolerable in a body of public knowledge that claims to be a science? This is the basic question in the philosophy of the social and behavioural sciences (§16.4), where it is almost impossible to get unambiguous data and to make unequivocal tests of theories.

8.6 Sociological epistemology

A sociological perspective not only throws light on the 'method' of science; it also illuminates the fundamental issues of scientific epistemology. The essential point is that many metaphysical concepts can be reinterpreted in social terms. Thus, for example, the notion of 'objectivity', which looms so large in the philosophy of science, really means no more than 'consensual intersubjectivity' (§3.2). There is no way of making an observation or arriving at an explanation that does not eventually entail human perception and cognition; this subjective element cannot be entirely eliminated, but it can be reduced to insignificance by reference to what is agreed on by different observers or thinkers faced with similar situations. It is misleading to postulate some non-human form of perception or cognition that comes into play when 'science' takes part in the game.

Other epistemological terms, such as 'belief' and 'intuition' seem to have reference primarily to the internal mental operations of individuals and belong, therefore, to what Popper has called 'world 2' (to distinguish it from the way things really are – 'world 1'); but this does not debar them from sociological analysis. In its most elementary form, the social model says little about the psychological attributes and activities of individual scientists. Nevertheless, their personal activities are not independent of their social environment. 'Public' knowledge stored in archives such as libraries ('world 3') has a major influence on the 'private' knowledge within individual minds. A consensual view is usually introjected into the mental worlds of most scientists in a particular generation, producing the resistance to paradigm change pointed out by Fleck and Kuhn (§7.3). Ordinary language, which is the medium of interpersonal communication and of individual thought, is common to both 'worlds'. Many well-known issues of social psychology and of sociolinguistics thus have a bearing on the sociology and philosophy of science.

Does this approach to metascience lead to an answer to the fundamental question of scientific epistemology – i.e. is science, so to speak, 'true'? Perhaps it does not. The notion of a scientific community of indefinite extent coming to a consensual opinion after critical debate is just as idealized as the traditional notion of the saintly scientist carrying out exemplary investigations and arriving at rational conclusions by irrefutable logic. A whole community can be just as wrong as any single individual, so we are no further ahead.

This model also fails to establish the 'reality' of the scientific world view, since there is no necessary connection between what science tells us (in 'world 3') and what things are really like – i.e. the presumed contents of 'world 1'. But it does go somewhat further than elementary *pragmatism*: it does say something more than 'science works, so for all practical purposes it is true'. The message is, rather 'science sometimes doesn't work, but in favourable circumstances it can be as true as *anything* is, in this mysterious world'.

The point is that radical epistemological relativism applies to *all* knowledge, including our most everyday perceptions of the 'life-world'. Instead of asserting the absolute truth and reality of this world, sociological *phenomenology* draws attention to the intersubjectivity of such representations, and the apparent objectivity this gives to them for all who share them. All human beings share an unshakable belief in the everyday world of spoons and sparrows, because we all find ourselves with each other in the same dream. In our society, at least, this is the firmest standard of factuality and 'truth'. In so far as science rests upon 'facts', and builds upon them, it may be taken as 'true'. The firmest epistemological basis for scientific knowledge is *empiricism* (§3.2). In the sociological formulation, just as in the traditional philosophical doctrine, the objectivity and truth of science cannot be any stronger than what we all hold in common about the everyday 'life-world' – but it need not be any weaker.

Further reading for chapter 8

A good basic text is
    M. Mulkay, *Science and the Sociology of Knowledge*. London: George Allen and Unwin, 1979

8.6 Sociological epistemology

The 'strong programme' is cogently formulated by
> D. Bloor, *Knowledge and Social Imagery*. London: Routledge & Kegan Paul, 1976 (pp. 1–46)

A study motivated by the 'strong programme' is
> H. M. Collins & T. J. Pinch, 'The Construction of the Paranormal: Nothing Unscientific is Happening'. In *Sociology of Scientific Knowledge*, ed. H. M. Collins, pp. 151–84. Bath: Bath University Press, 1982

Some of the arguments against the sociology of knowledge are given by
> L. Laudan, *Progress and its Problems*. Berkeley, Calif.: University of California Press, 1977 (pp. 196–222)

The general case for radical epistemological relativism is made by
> B. Barnes, *T. S. Kuhn and Social Science*. New York: Columbia University Press, 1982

A sociological epistemology of science is developed by
> J. Ziman, *Reliable Knowledge*. Cambridge: Cambridge University Press, 1979

The notion of 'world 3' is to be found in
> K. Popper, *Objective Knowledge*. Oxford: Clarendon Press, 1972 (pp. 126–8)

9

Science and technology

'The outlook for gaining useful energy from the atoms by artificial processes of transformation does not look very promising.'   *Ernest Rutherford (1937)*

9.1 Science as an instrument

Up to this point, we have been looking at science from the 'inside': now we take an entirely different standpoint, and consider science from the 'outside'. The *external sociology* of science considers it simply as a social institution, embedded in society, and performing certain functions for society as a whole, on a par with other institutions associated with law, religion, political authority and so on. For the moment we shall treat science as a 'black box', whose inner workings are of no significance except to ensure that it can perform the functions assigned to it. Eventually (§12.5) we shall reopen this box, and reconsider the internal sociology and philosophy of science from an externalist point of view.

Science is valued by ordinary citizens, by powerful individuals such as politicians and company directors, and by corporate bodies such as commercial firms and government agencies, primarily for its *use*. It is fostered mainly as a resource to be applied to the furtherance of individual and/or collective activities whose goals are *not specifically the advancement of knowledge*. This conception of science as essentially an *instrument* for achieving a variety of goals *other* than the acquisition of knowledge is so widespread and so dominant in our society that it overshadows all other conceptions of its social function. It thus takes no account of several attributes of science that often motivate scientists personally, such as the religious gratification of 'making manifest the handiwork of God, as revealed in nature', or the aesthetic satisfaction of 'discovering and explaining the marvels and mysteries of the world about us'. The instrumental conception of science also ignores the *reflexivity* of the relationship between society and science, which not only transforms its material basis but is also a major element in its ideological superstructure (to use the Marxist terminology). The 'scientific world view' is so inextricably woven into the fabric of modern social, political, religious and aesthetic thought that science cannot be treated simply as the means for reaching ends that have been chosen on 'non-scientific' grounds. These issues will be discussed in chapter 16.

This conception of science, which justifies research primarily as a medium of conscious social action, dates back at least to Francis Bacon, at the beginning of the seventeenth century. Since that time, science has been deliberately fostered, supported, financed and planned, by individuals, by corporate bodies, and by the state, on an ever-increasing scale. The dominant theme in the social and economic history of 'Western Civilization' has been the growing influence of science, in all its multifarious modes. But the effect on society of the various branches of science, and of their associated technologies in medicine, engineering, agriculture, etc., is too large a subject to be dealt with historically in this book. Let us consider them therefore very schematically, by their degree of concentration of use, in a contemporary setting.

9.2 Science-based technology

The most striking influence of science on society is the generation of an essentially novel *technology* out of basic, discovery-oriented research. The prime example is the *electrical industry*, which grew up in the late nineteenth century as a direct outcome of the pioneering researches of Michael Faraday – and many others – in the early part of the century. The development of this industry by inventors and entrepreneurs such as Thomas Edison and Werner Siemens cannot be imagined without the theoretical understanding and empirical knowledge obtained previously by 'pure' scientists who had no direct utilitarian motives.

A twentieth-century example is the development of *nuclear engineering*, both for weapons and for electrical power generation. This gigantic technology, is based directly upon the primary researches of academic scientists, such as Ernest Rutherford and Enrico Fermi, which were undertaken in the firm belief that their discoveries were most unlikely to be put to any important practical use. A development that is now under way, with unforeseeable consequences for the twenty-first century, is the application of fundamental understanding of the molecular basis of heredity to industrial and medical ends, in the form of *biotechnology*.

Quite novel *science-based technologies* may be generated from basic science on a variety of scales. Thus *radar* developed out of academic research on the propagation of radio waves in the upper atmosphere of the Earth, whilst the principle of the *laser* was derived from the fundamental theories required to explain quantum phenomena in atoms. It is a commonplace of modern engineering, medicine and agriculture that completely novel techniques and devices may be conceived and put to use by the exercise of scientific knowledge which was originally acquired 'for its own sake', or in the pursuit of quite different ends. Thus, the knowledge that accumulates in the scientific archives (§4.1) can be considered a vast resource to be exploited for its unsuspected technological uses.

9.3 Technology-based sciences

It is important to realize that not all advanced technologies derive from basic science. Thus, for example, the practical techniques of *mining* and *metallurgy* have their origins in the mists of antiquity, and continue to be extended and improved by inventive craftsmanship and imaginative design. Most of the patentable inventions incorporated into the design and manufacture of a modern motor car were produced in this way, by workshop engineers rather than by laboratory scientists.

But many traditional techniques (§16.3) have proved amenable to scientific study, and have been found to have an underlying scientific rationale. This applies particularly to *medicine*, whose therapeutic arts have been studied systematically from the time of the Ancient Greeks. The effort to understand and master the natural phenomena of human disease has thus developed into a highly sophisticated science, with a characteristic body of deep theory to explain these phenomena and bring them under control. In a similar way, a variety of ancient crafts were transformed in the nineteenth century into the *technology-based science* of *industrial chemistry*, whilst in the twentieth century the practical technical knowledge of the metallurgist has been incorporated in a new science of *materials*. The same process is to be observed in almost all fields of practical human activity: 'technologies' such as agriculture, civil engineering, food processing, architecture, etc., have developed their respective 'sciences' to guide further technical progress.

9.4 Scientific technique

Quite apart from their specific applications in advanced technologies, the ideas, concepts, theories, instruments, data and techniques of science permeate practical life. Inventors, farmers, parents, motor mechanics, builders and other persons in innumerable skilled and semi-skilled professions acquire a rough outline of the scientific views of the day, and apply them artlessly to the solution of day-to-day problems. Thus, for example, the concept of *energy*, around which the science of thermodynamics was developed in the mid-nineteenth century, is the key variable in every practical decision concerning fuel resources, power generation, space heating, propulsive efficiency of vehicles, etc. Again, biochemistry and physiology have provided the basic facts and theories of the practical science of *nutrition*, so that 'everybody' nowadays knows about calories and vitamins, and tries to act in accordance with this knowledge.

These applications of science are so pervasive and so intangible that they are often overlooked. The fact is, however, that people not only make use of the products of science-based technologies, such as pocket calculators and pep-pills; they also use elementary science-based techniques in dealing with practical problems, and orient themselves in the life-world by science-based modes of thought (§16.3). Where these techniques and modes of thought are lacking, as they still are amongst the general population of most developing countries, the instrumental function of science in society is greatly reduced. Thus, for example, complete ignorance of the bacterial causes of disease is one of the main obstacles to the widespread use of scientific methods of elementary hygiene in many countries. At this point, of course, we are not saying whether the influence of modern science and technology in the Third World is good or bad; we are merely noting that this influence is not to be measured solely in terms of rice yields and expenditure on machine guns.

9.5 *Science* or *technology*

One of the most tangled issues in the study of science and technology is the relationship between these two terms. It is easy to give clearcut examples of each category, such as cosmology on the one hand and automobile manufacture on the other, but where to do we draw the line between them? Until recently it was customary to make a distinction between science as the generation of knowledge primarily for its own sake, and technology as a body of knowledge concerning a practical technique. Unfortunately this convenient distinction has not been maintained in common use, where a decision to build a computer factory is described as *science* policy, and the computer itself is called a piece of modern *technology*. For this reason, the term *academic* science was used in previous chapters, to indicate that the discussion was primarily about science in that traditional sense.

But the difficulty is not purely semantic. In its strict meaning as a body of *knowledge* concerning a technique, rather than the routine practice of the technique or its material products, every technology is committed to the regulative principles of 'science' (§3.10). Whether this knowledge can be regarded as scientific then depends upon one's notion of what further criteria must be satisfied. Must it be theoretically explanatory and predictive, for example, as a philosopher might insist, or available in a public archive as a sociologist might argue? Or should the distinction still rest on the purpose for which the knowledge is sought?

Historically speaking, every technology tends to become more and more subject to the characteristic 'method' of science. A practical craft, such as pottery or ploughing, may have been passed on from generation to generation, by imitative apprenticeship with very little formal instruction. Although this process may permit a subtle and sophisticated evolution of the tacit knowledge embodied in the craft (§3.3), it still lacks the explicitness and generality of a genuine science. But any attempt to codify this knowledge into a course of instruction or a handbook makes it explicit and forces it into a categorial framework. General principles are invoked to justify a classification scheme where patterns of cause and effect (§2.11) (say) become apparent and call for explanation. It is a short step, then, to framing hypotheses (§2.14) and testing them experimentally (§3.7), in the search for an overall theoretical description of the various phenomena observed. In other words, information concerning the technique becomes the domain of a fully fledged science, where research is undertaken and knowledge validated according to the same epistemological principles as in more academic disciplines.

An advanced technology, such as aeronautical engineering, is to be distinguished from the associated science of, say, aerodynamics, by the inclusion of a great deal of empirical 'know-how' which is so far from rational explanation that it almost defies codification. Much of this knowledge is tacit, or is kept secret, for commercial or military reasons, so that it is not available in archival form. Nevertheless, there is an historical tendency for all crafts to become codified as technologies, and for all technologies to give birth to regular sciences intended to bring the craft under predictive control. This tendency is clearly one of the most significant instrumental characteristics of science in modern society.

9.6 Science from technology?

An immediate proposition suggests itself: perhaps *all* science is simply an intensified form of technology, generated by the material needs of society. This has been a major contention of Marxist theory, ever since it was proposed unequivocally by Boris Hessen at a famous meeting in London in 1931. This thesis is closely connected, of course, with the general body of Marxist thought, and cannot be discussed in full without reference to the whole conceptual apparatus of dialectic materialism and the rôle of science and technology in the class struggle. But it can be treated as a hypothesis worthy of empirical test.

The standard case in favour of the Hessen thesis is the history of the *steam engine*. This immensely influential technology was developed from the late seventeenth century until about the middle of the nineteenth-century by essentially practical men, using the traditional craft skills of the mechanical engineer. Although this development was undoubtedly indebted to academic science for a few key ideas, such as 'the power of the vacuum', and the latent heat of steam, it was mostly carried out by trial and error, in the light of day-to-day experience, without recourse to abstract analysis. These were men very close to the material, technical base of the society of their day, and were simply responding to the commercial need for a means of pumping water out of deepening mines. The capitalist entrepreneurs who fostered this development were not in the least interested in science; they asked only for profitability and

9.6 Science from technology

efficiency – how little coal need be burnt to pump such and such a quantity of water from such and such a depth, and how would the cost of this coal compare with the cost of feeding horses to do the same amount of work?

By the first half of the nineteenth century, steam engineering was a mature, well-codified technology and was thus a natural subject for experimental investigation and theoretical analysis. The work of Sadi Carnot, James Prescott Joule, William Thomson and many others created a new science of *thermodynamics* which not only gave an exact quantitative account of the behaviour of all heat engines, but also incorporated many of the basic principles of academic physics, including Newton's famous laws of motion. A new general theory had thus been conceived, motivated by economic demand, drawing thematically (§2.14) upon quasi-economic analogies, and validated pragmatically by its technological achievements. From this time onwards, thermodynamics was available as a primary theoretical discipline for the design of new industrial products to fulfil new material and commercial needs – the internal combustion engine for road vehicles, the steam turbine for electric power generation and for ship propulsion, and eventually the turbojet engine for military and civil aircraft. But it was more than a resource for technological innovation: the laws of thermodynamics were reformulated in abstract form, and became the basis for new branches of academic science such as low-temperature physics, physical chemistry and meteorology. Thus we may say that a considerable proportion of our present understanding of the natural world can be traced back to the desperate need for some means of pumping water out of mines, and thus maintaining the profitability of a highly capitalized industry.

This case history of a technology-based science, which can be paralleled in other fields of engineering, agriculture and medicine, is good evidence in support of the Hessen thesis. But this thesis completely fails to explain the development of science-based technologies (§9.2), such as the electrical and nuclear power industries, which did not grow out of pre-existing techniques, and were not generated by research and invention directed towards meeting a perceived need. No amount of commercial demand for a means of transmitting information and energy instantaneously to a great distance, or military demand for an explosive that would destroy a whole city, could have produced these technologies before the discovery of the scientific principles on which they were later based – and it is quite clear from the historical record that the scientists who made these discoveries did not have these applications in mind.

Indeed, the characteristics of most science-based technologies indicate quite a different model for the social rôle of science. Such technologies are often fundamentally *innovative* in that they evolve into the means of attaining technical goals that were previously regarded as quite impossible to approach except by magic. Imagine, for example, what people would think of the idea of transmitting speech instantaneously to the other side of the world, before the invention of the electric telegraph and telephone. These capabilities are not only unprecedented; they are also *unpredictable* in principle, since they do not arise by the imaginative extrapolation of existing techniques but by the exploitation of apparently irrelevant discoveries. In many cases, as with the discovery of X-rays, this discovery itself may be serendipitous (§2.5) even within its original scientific context.

These characteristics make such technologies profoundly revolutionary, and yet beyond conscious control. There is no need to emphasize the extent to which they have transformed the everyday life and means of production in all advanced industrial societies – a transformation that eventually extends to the political and social structure of those societies. But the control over nature, and over other people, that can be exercised by means of advanced technologies does not apply to the evolution of those technologies themselves. A ruling class may try to appropriate the applications of basic science embodied in such instruments of authority as a television set or a guided missile, but it has no means of directing or foreseeing future discoveries which may radically change its own position (§14.5). Various sciences may acquire unchallengeable paradigms (§7.3) on which to base mature and efficacious technologies, but the notion of *finalization* (§12.3) – the deliberate choice of the ends to be achieved by the further pursuit of these sciences – is an illusion. Thus, paradoxically, the greater the certainty and power with which known scientific technologies can be applied to existing situations, the greater the uncertainty and sense of powerlessness that scientific progress introduces into social and political affairs over a longer term.

In contrast, therefore, to the thesis that science should be considered subordinate to social and political forces, there is a well-founded view that it is an *autonomous* factor in society, capable of producing immense changes which could not be predicted solely in terms of the interests of economic classes, entrenched institutions, or other conventional political agencies. This factor is so indeterminate over a period of a few decades that it makes nonsense of all attempts to foresee – and to try to forestall – the course of history. On this view, a creative social rôle for science can only be accommodated in a pluralistic model of society which repudiates all historicist claims.

9.7 'S & T'

The Marxist and pluralistic accounts of the origins of science and its relationship with technology carry conflicting political and social implications, which will keep appearing as we proceed further into the external sociology of science. This was, for example, the underlying theme of the public controversy on 'freedom in science' that took place in Britain in the 1930s. But this polarization and direct confrontation along political lines is too simplistic, since it bears no relation to the way in which things go in practice. Historically speaking, we observe distinct cases of both 'science-based technologies' and 'technology-based sciences' – and a variety of intermediate cases where technological demand has had a more or less important influence on the evolution of an academic scientific discipline.

In reality, these categories merge into one another, and confound most of the distinctions between a 'science' and a 'technology'. Should one really distinguish between the *steel* industry and the *polymer* industry because the former has an ancient craft tradition? Is *nuclear physics* less practical and less socially relevant than *hydrodynamics* because the latter has roots in hydraulic engineering and shipbuilding? Is *aeronautical engineering* essentially more scientific than *architecture* because it makes more deliberate use of recent scientific discoveries and research methods? Was *pharmacology* not really a science until the last few years because it had not acquired the paradigms of molecular biology and relied heavily on trial and error following on chance discovery? The more one considers such questions, the more one appreciates their futility.

It is difficult nowadays to find any material activity of society that does not turn to the production of knowledge by research as a means of achieving its particular goals. Thus, all technologies are in the process of generating their respective sciences. Conversely, it is difficult to find any body of knowledge, however derived, that is not being scrutinized for its potential benefits in material form. Thus all sciences are in the process of generating their respective technologies. These processes are intermingled on every scale, from the laboratory and workshop to the research council and industrial firm, and in every dimension of interpenetration. We observe the growth of hybrid institutions, such as the 'Science and Engineering Research Council', hybrid techniques such as electron microscopy, and hybrid disciplines such as clinical neurophysiology.

The rôle of science in society is thus inseparable from the rôle of technology. These are merely two aspects of an indivisible activity: *Science and Technology* – 'S & T'. This activity is embodied in a variety of social institutions whose primary function is essentially instrumental. In the short or the long run, they justify their existence by producing *practical* knowledge, in the form of designs for novel, humanly relevant products or techniques that can be put to some use. These uses range over a very wide ethical spectrum, from meeting basic human needs for food, shelter and health to supporting the power structure of society with warlike weapons and profitable investments. But this function can only be carried out effectively by the generation of an intermediate product – generalized or 'academic' knowledge – which is not immediately practical. As we shall see in chapter 10, some S & T organizations specialize in the production of this type of knowledge, whilst others are mainly involved in transforming it into practical forms. But these are not separate social rôles, and are often performed simultaneously by the same people in the same organizations (§12.2). This may not be how science and technology have appeared in the past, but this is the way they now look from the standpoint of society at large.

Further reading for chapter 9

The classic text on the instrumental conception of science is
> J. D. Bernal, *The Social Function of Science*. London: Routledge, 1939 (especially pp. 1–34)

A history of science that emphasizes the technological connection is
> J. D. Bernal, *Science in History*. London: Watts, 1954

The controversy surrounding the Hessen thesis is described by
> G. Werskey, *The Visible College*. London: Allen Lane, 1978 (pp. 139ff and 181ff)

A Marxist interpretation of the social rôle of science is presented by
> H. Rose & S. Rose, 'The Incorporation of Science', in *The Political Economy of Science*, ed. H. Rose & S. Rose, pp. 14–31. London: Macmillan, 1976

Some recent papers on the relation between science and technology include
> O. Mayr, 'The Science–Technology Relationship' and D. J. de S. Price, 'The Parallel Structures of Science and Technology', in *Science in Context*, ed. B. Barnes & D. Edge, pp. 155–63, 164–76. Milton Keynes: The Open University Press, 1982

and
> G. Böhme, W. van den Daele & W. Krohn, 'The "Scientification" of Technology' (pp. 219–50) and P. Weingart, 'The Relation between Science and Technology' (pp. 251–86); in *The Dynamics of Science and Technology*, ed. W. Krohn, E. T. Layton & P. Weingart, Dordrecht: D. Reidel, 1978

10

Pure and applied science

'The constant activity which you Venetians display in your famous Arsenal suggests to the studious mind a large field for investigation.' *Galileo Galilei*

10.1 'R & D' in 'S & T'

Science and Technology – perhaps one should say, 'the sciences and their associated technologies' – together constitute a major social institution based upon the systematic generation, accumulation and utilization of knowledge. This knowledge is very diverse. Some of it is directly useful; some of it appears totally divorced from human affairs. Some of it is symbolically codified in the form of experimental data, theoretical formulae, solutions to standard problems, therapeutic protocols and engineering blueprints: some of it is essentially tacit, and only becomes manifest through expert technical work (§15.4). Much of the knowledge that is put to use has simply accumulated in the scientific and technical archives, over a period of many years. As in the past, a considerable amount of formal technological knowledge is continually being produced in day-to-day practice; in clinical medicine, for example, any novel course of treatment may be considered something of an experiment.

The immense social dynamism of modern 'S & T' comes from its aggressive employment of the social device of *research* – that is to say, *systematic activity undertaken to obtain information or understanding that goes beyond established knowledge or accepted practice*. This notion is, of course, very familiar psychologically and philosophically as the intentional factor in the process of scientific discovery (chapter 2). We should now come to see it sociologically as a peculiarly powerful means for – as the founders of the Royal Society put it – 'effecting all things possible'.

In contemporary discourse, the notion of scientific 'research' is usually closely linked with the notion of technological 'development'. The relationship between these two terms mirrors the relationship between 'science' and 'technology'. Until recently, there were considerable cultural differences between philosophical speculation and experimentation, on the one hand, and technical invention and innovation on the other. It is true that both these activities might have been carried out in the same general spirit of imaginative enterprise and sceptical rationality. The essential characteristics of both forms of activity can be detected far back in antiquity in all the major civilizations of the world, and as the history of Chinese science and technology shows, the boundaries between them are very elastic. But in the European tradition it has been customary to distinguish between a concept of 'research', designed simply to produce information or understanding, and a concept of 'development' designed to produce a useful object or process out of a novel idea.

Like the distinction between science and technology (§9.7), the distinction between 'research' and 'development' is becoming more and more difficult to discern. The combination 'Research and Development' – 'R & D' – can be used to cover a whole range of investigatory procedures, from the most abstract theoretical analysis to the most earthy trial and error, which may be found in all fields of 'S & T', whether motivated by a desire for fundamental understanding, as in cosmology, or by a desire to make a commercial profit, as in the manufacture of weapon systems. It is often convenient to assess the degree of *relevance* of a particular R & D project or programme (§12.2), but as we shall see in later chapters this is a continuous variable, spread across a broad spectrum, and is not always a reliable indicator of the character of the work being done or the nature of the organization within which it is carried out.

A striking characteristic of R & D is that, like modern S & T, it exists in a single world-wide mode, closely associated with the European style of life. This mode is linked, of course, with European technological and commercial dominance in world affairs in the nineteenth and twentieth centuries, but is much less 'technological' than 'scientific' in spirit. It is derived to a remarkable extent from the academic style of science that developed in Western Europe from the late seventeenth century onwards and was fully institutionalized by the end of the nineteenth century. For this reason, the internal sociology of academic science, to which we have given so much attention in earlier chapters, is highly relevant to the external sociology of contemporary S & T, even though the academic style itself is being rapidly transformed – may already be obsolete – under the pressure of socioeconomic forces (§12.5). Psychologically, philosophically, and at least microsociologically, 'academic research' provides rôle models, validating principles, ideological traditions and behavioural norms for the whole R & D process, whether in an educational institution or in a commercial firm.

This is not to assert that R & D is just academic research organized for use. On the contrary, the institutional forms of 'industrial' science (§10.6) are quite different from the traditional forms of 'academic science'. Modern R & D organizations (chapter 12) incorporate fundamental features drawn from professional practice, as in medicine; from the art of design, as in engineering; from economic pragmatism, as in industrial invention and production (§13.4); and from administrative rationality, as in governmental science (§14.4). These features are growing in significance, but they do not yet predominate over the 'scientific research' theme in all R & D. It is necessary, therefore, to look briefly at the social history of academic science (considered, for the moment, as a distinct institution) in order to understand something of how this institution was actually related to its social context.

10.2 Growth

The history of science is dominated by *growth*. The internal sociology of science is not only adapted to cognitive and institutional change (§7.1): every scientific community has had to accommodate increasing numbers of scientists at a rate that far outstrips the growth of the population as a whole. This growth has not, of course, been historically or geographically uniform, but the scale factor for the growth of the scientific literature – something like 100-fold per century (equivalent to a doubling time of about 15 years) – gives an idea of its magnitude. In other words, for every scientist in the world in the latter half of the seventeenth century, there are now about a million. Even if we discount this estimate by a factor of ten, to allow for changing notions of the scientific enterprise, it remains a phenomenal figure.

Science is thus seen by everyone to be an immensely *successful* social activity, which has gone on from strength to strength, with scarcely any check. This continuing success goes far to explain its prestige in society at large, and the extent to which it is linked optimistically to the ideology of progress (§3.10). The buoyant internal morale of an expanding enterprise is an influential factor in its external relations.

The size of the scientific community is also an indicator of its weight and influence in society at large. Science, in the broad sense of the term, now involves, directly or indirectly, something like 1% of the population and of the national income of most advanced industrial countries, and is therefore a significant factor in demographic and economic affairs. Whether or not this represents a saturation level in personnel and other resources, it is a minimal indication of the relative importance of science in modern society. The history of science thus tells of a long smooth transition from a quite insignificant to a major social institution.

10.3 Amateurism and state patronage

Until quite recent times very few scientists were employed specifically to do research. This *amateur* status included those who held posts in universities, where their function was primarily to teach what was already well established rather than to undertake original investigations on their own account. A professorial appointment was, of course, attractive to a man of scholarly accomplishment, and provided access to higher learning and the leisure for research. But scientists also got their living from several other learned professions such as medicine, the law, or the church. Many 'savants', indeed, were people with private incomes from land or commerce, whose diligence in research was sustained solely by personal enthusiasm and the recognition of their peers. Although some of the most famous early scientists, such as Hooke and van Leeuwenhoek, were of quite lowly origins, research was regarded as essentially a gentlemanly pursuit at which one could scarcely hope to make a living. This is evident in the constitution and membership of communal scientific institutions, such as the Royal Society, which were quite clearly genteel – even aristocratic – in their class status.

In Britain, science remained almost entirely outside the state apparatus until the late nineteenth century. But in more authoritarian European countries such as France, Prussia and Russia, science acquired considerable official patronage, especially through the establishment of 'academies' with paid posts for a small number of eminent scholars. In the era of the nation-state, scientific achievement became a competitive resource for patriotic pride, and scientists of renown were formally enrolled as State functionaries, although often without any specific governmental function or bureaucratic authority. At various periods in various countries scientific *ideas* have been considered very radical and subversive by a ruling élite, but science has seldom been *institutionally* antipathetic to, or incompatible with, the state apparatus or other major instruments of power. Despite their international connections and cosmopolitan careers, natural scientists could almost always be counted on for their loyalty to their country and class.

10.4 The rise of academic science

The most significant phase in the institutional history of science took place in the first half of the nineteenth century, when science began to move into the universities. This change was most marked in Germany, and on a smaller scale in France. The *academicization* of science in Britain and the United States followed somewhat later, and largely copied German models.

The rise of the German universities to supremacy in the European academic world was a complex historical process with numerous contributory causes. It is enough to remark that Germany was then divided into a number of quasi-independent states, which could compete publicly through the scholarly reputation of their local state universities. On the other hand, teachers and students could move fairly freely throughout a large region where a single language was widely spoken and where there was a relatively homogeneous culture – a region that effectively included Switzerland, Holland, Scandinavia, and parts of what are now separate East European countries such as Poland, Czechoslovakia and Hungary. Under these conditions, scientific research took on a new professional and institutional rôle in society.

In the first place, all research was sited in the university; it was carried out on the campus, by university employees. But the principal researchers were *professors*, who were primarily employed as university *teachers*, and were not directly answerable to any authority for the intensity, intentions, or outcome of their scientific activities. They might, indeed, do no research at all, but if they did the results were publishable without constraint; the general principle of academic freedom was regarded as paramount.

Entry into this profession was, however, stringently competitive. After the usual period of undergraduate study, it was necessary to become apprenticed (without pay!) to an established researcher, and to undertake the research on which to write a scholarly dissertation. There might then be a further period of years in a poorly paid subordinate position as a teacher/researcher before one eventually won a professorial appointment with a good salary and permanent tenure.

This was, in essence, the career structure of German academic science, which can easily be recognized as the prototype of the career structure of all university systems throughout the world in the twentieth century. It is peculiarly ambivalent in the social rôle it assigns to research performance, which is not contracted for as such, but which is the indispensable criterion for academic employment and preferment. Indeed, this ambivalence is so extreme that universities are quite accustomed to giving 'lectureships' to people who can scarcely put two sentences together on a public rostrum, provided that they have a few really good scientific papers to their name!

It scarcely needs to be repeated (cf. §6.1) that this is the social mechanism that really enforces the authority structure (chapter 5) and behavioural norms (chapter 6) of academic science. 'Recognition' by promotion to a full professorship, with permanent tenure, is as powerful a material incentive for quality and quantity of research production as any amount of ego gratification by a symbolic token of esteem such as an honorary degree. But it is not, like most bureaucratic systems, self-justifying and self-sustaining. This way of making educational appointments presupposes the existence of an autonomous scientific community where people can win genuine personal reputations for their research. It also implies that the individualism of the traditional academic ethos (§5.2) should not be too narrowly limited by institutional constraints; one of the significant characteristics of the German universities was that they were run by a professorial oligarchy, with remarkably little interference from the State apparatus to which they were, in principle, answerable.

10.5 The external relations of academic science

In thinking about science, it is essential not to disconnect the philosophy from the sociology of the scientific community: it is equally necessary not to treat the scientific community as if it were quite independent of its material base. The ethos of academic science is so treasured (§6.4), and so epistemologically fruitful (§8.4), that we may forget that it is an ideology that fits very closely the peculiar interests of a particular occupational group. The external sociology of science must take account of the way that academic scientists were able to maintain a remarkable degree of autonomy in their research, despite the fact that they were paid employees and responsible officials of institutions of higher education, which are firmly embedded in the affairs of society at large. In return for their work as teachers, they were permitted – indeed strongly encouraged – to 'push back the frontiers of knowledge' with little reference to the needs of commerce or the State. Science was thus established as a recognized profession, without the direct social responsibilities of most other professions, which are more immediately involved in worldly affairs.

But academic science was never entirely disconnected from the larger society. Academic authorities were called in to advise their governments on technical questions (§15.5) and also played an important part in the management of various governmental agencies for surveying, astronomical observation, botanical collection, etc. (§14.6). In their turn, academics would seek state support for major research enterprises, such as a geographical expedition or an international magnetic survey (§14.4).

Scientific knowledge was, of course, continually flowing outward, through scientific publications and scientific education, to industry and commerce. In some disciplines, such as medicine and engineering, the academic researchers were also élite professional practitioners, and thus directly involved in everyday life and in public policy. As teachers they were also responsible for the training and licensing of the mass of practitioners, who needed to learn the latest scientific methods. Although there were few formal links between academia and industry, some academic scientists of high standing – for example Lord Kelvin – would be called in as consultants on technical matters, and most scientists were personally interested in current technological developments. Although it held itself somewhat aloof from everyday concerns, the scientific community was not unaware of the value of good public relations (§16.2). Many distinguished scientists dutifully performed as popularizers – if not propagandists – of science, through such public media as the British and American Associations for the Advancement of Science. Science was presented as a highly respectable profession, deserving its privileged place in society by its contributions to progress and prosperity.

This stereotyped societal rôle for academic science had evolved in Germany by about 1850. It soon spread to other countries, and remained effective for something like a century. There were, of course, local variants on this stereotype, depending on local cultural, political and economic circumstances. Nevertheless, in spite of the immense changes of scale and of intellectual style produced by the growth of science, of industry, and of higher education, in that period, the academic model of scientific work has proved remarkably durable. This model evidently combines epistemological, occupational and societal functions in a stable, self-consistent way. To what extent does this stability depend upon the nature of the social system as a whole? There is undoubtedly a close connection between the academic ethos and the ideology of a pluralistic capitalist democracy (§6.4). At all events, academic science is clearly a coherent institutional form, capable of maintaining an independent existence for a long period in any reasonably open, economically developed society.

10.6 Industrial science

Even in its heyday, in the early twentieth century, academic science was not the only institutional model for research. From the middle of the nineteenth century onwards, there has developed an alternative model in which scientific workers were employed directly, on a full-time basis, to do research. Firms in advanced industries such as chemical manufacture had, of course, always benefited from scientific discoveries, and often employed people with a scientific training as works managers or process controllers. But in the 1860's the German dyestuffs manufacturers took a decisive step forward by setting up their own company laboratories, where fully qualified academic scientists were employed to undertake independent investigations in the hope of discovering new products and processes. *Industrial science* was thus established as a major instrument of innovation in all science-based industries (§9.2), such as chemical engineering, electronics and aeronautical engineering.

A parallel development also took place in the various government agencies, such as astronomical observatories, geological surveys, bureaux of weights and measures, and public health inspectorates, which provided routine technical services for the public benefit. The work of these organizations was usually based upon scientific principles, and was carried out by people with scientific qualifications: when they were faced with novel technical problems, or areas of basic ignorance, they naturally moved towards a research attitude and acquired specific research functions. Although this sort of *governmental science* differed in many details from more commercially oriented industrial research, there were such structural similarities between, say, the National Physical Laboratory on the one hand, and the General Electric Company's laboratory on the other, that the term 'industrial' may be applied to both.

The most striking characteristics of industrial science as an institutional form were those in which it differed from academic science. Its establishments were not normally sited in universities, and its staff members had no direct educational responsibilities. A typical industrial research laboratory was not a quasi-autonomous organization, but was usually a bureaucratic sub-division of some much larger non-scientific organization, such as an industrial firm or government department. Individual staff scientists, however senior, were not free to follow their own bent in the choice of research projects, but were expected – and at times firmly directed – to work towards the goals of the superior organization. Their duty was to invent a new commercial product, map a specific area of the country, or perfect a new technique of measurement, not simply to acquire knowledge. In any case, the results of their research had to be put at the disposal of the employing organization, and might even be kept secret for commercial or military reasons.

Since the training of research workers was not a specific function of industrial science, employees were recruited from academia on completion of a bachelor's or doctor's degree. Their subsequent career would depend more on local organizational considerations, such as the quality of their technical performance, or managerial competence, than on public reputation within the scientific community, and was subject to the same administrative regulations and management decisions as any other employees of the firm or government. In fact the 'laboratory' or 'establishment' as a whole would be structured internally, and directed, according to the standard procedures of the parent body – that is, through a bureaucratic hierarchy of 'sections' and 'divisions' up to the very top.

From this brief sketch, it is obvious that industrial science was very different from academic science as an institutional form. It had a different internal sociology, different incentives and rewards for the individual, and a different rôle in society. It was scarcely to be considered a distinct social institution, associated with an autonomous 'community', but mainly derived from, and referred to, its various parent institutions in society at large. In other words, although it embodied the scientific notion of 'research', and drew heavily on the contents of academic science, it was designed around the instrumental conception of science as a means of achieving particular practical ends (§9.1) with little reference to the conception of science as a process of discovery.

By the beginning of the Second World War, industrial science had developed into a distinct way of life (§15.4), embodied in such stable and successful institutions as the Bell Telephone Laboratories or the Royal Aircraft Establishment, which employed a considerable proportion of the scientists, and played an indispensable part in the economic affairs, of all industrialized countries. As we shall see in the next two chapters, this way of life is becoming the dominant form in the 'collectivized' science of our times.

10.7 Pure science — and its applications

The distinction between the academic and industrial modes of research is often strongly emphasized in any account of the societal function of science. Until the last few years, this distinction was not merely institutional; it was even reinforced by status symbols, such as membership of learned societies. In Britain, for example, academic physicists would belong to the Physical Society, which published a learned journal and rewarded its most esteemed members with prizes; industrial physicists joined the Institute of Physics, which was concerned with professional qualifications and conditions of employment. In other countries, where engineering was not so stigmatized, this snobbish differentiation may not have been so pronounced, but the distinction in principle between the two institutional models was seldom forgotten.

In reality, this distinction was not as sharp as many people believed. Thus, for example, they were not differentiated epistemologically or methodologically; industrial science used just the same theories, methods, concepts and terminology as academic science. Industrial scientists went through the same educational institutions as academic scientists, and often had experience of academic research at quite a high level of seniority. Industrial scientists often sought recognition from the academic community, and published papers were the primary criterion for promotion in many branches of the scientific civil service. Indeed, in disciplines such as agriculture the research was mainly done in hybrid organizations, where scientists were relatively free to follow the norms of academic science within a formally bureaucratic framework.

Nevertheless, in spite of the radical transformation of the social relations of science and technology in the past few decades, a distinction between 'pure' and 'applied' science still lingers on. This distinction is not at all the same as between the 'research' and 'development' components of 'R & D', since the work of an 'applied scientist' in an industrial laboratory might be directed as much towards the explanation of general phenomena, or the determination of basic data, as towards testing and improving a specific product or design. Nor can applied science be simply equated with 'technology', which almost always contains a large ingredient of tacit knowledge and traditional craft lore (§9.5). Although engineers make use of a great deal of scientific knowledge, they are not just 'applied scientists' in their professional work.

As the epithet 'pure' suggests, this distinction is essentially ideological. It asserts the independence of academic science from all material or social considerations, and proclaims the virtue of doing research 'for its own sake'. It repudiates the instrumental conception of science (§9.1), and thus preserves the academic ethos (§6.3). The implicit argument is that although the ultimate social value of science comes through its applications, these are unforeseeable (§9.6), and must not in any way influence the process of scientific discovery, which follows its own peculiar laws. In other words, according to this ideology, the internal sociology of the scientific community and the external sociology of 'science and technology' are to be considered quite separate topics for metascientific study.

Sociologically speaking, however, the notion of an essential difference between 'pure' science and 'applied' science is difficult to sustain. This notion may well have been justified half a century or so ago, when academic and industrial institutions were kept well apart, but it is no longer a valid distinction in the collectivized R & D system of today (see chapters 11 and 12). The institutional forms, the internal sociology and the societal relations of the elements of this system can no longer be classified by their position along the traditional axis from pure science to its applications.

In any case, it was never philosophically or psychologically convincing to insist that the fundamental character of research depends on the supposed purpose for which it is undertaken. In practice, science is mostly 'problem-solving' (§2.15), and it usually makes very little difference whether the problem to be solved is a question arising out of the paradigmatic research programme of an academic discipline (§7.3) or whether it is chosen because it happens to be relevant to some practical human need. If the question were, for example, how the roots of plants absorb minerals from the soil, would it make any significant difference to the method of investigation or to the validity of a discovery claim, whether or not this particular problem was being studied because it might have some significance for the use of artificial fertilizers or the growth of crops on saline soils? At the level of 'laboratory life', straightforward methodological and conceptual considerations determine the nature of the research process and the attitudes of those involved in it, whether it is as 'pure' as cosmology or as 'applied' as the search for a cure for the common cold.

It is worth remarking, nevertheless, that these methodological and conceptual considerations are almost invariably drawn from the academic tradition of science, even when the research is essentially 'applied', and the institutional framework is essentially 'industrial'. Technological development itself has become 'scientific': it is no longer satisfactory, in the design of a new automobile, say, to rely on rule of thumb, cut and fit, or simple trial and error. Data are collected, phenomena are observed, hypotheses are proposed, and theories are tested in the true spirit of the hypothetico-deductive method (§3.7). The R & D process of modern S & T is regulated by just the same principles of scientific work (§3.10) as those that presided over the growth of the natural sciences themselves. Epistemologically speaking, all S & T takes its cues from academic science, and is therefore steeped in its rôle models, its ethos and its institutional traditions. That is why we keep coming back to the internal sociology of this form of science in all our efforts to understand science and technology in their social context.

Further reading for chapter 10

Many of the works recommended for chapter 9 contain material relevant to the theme of the present chapter.

A classic work on the societal rôle of the scientist, which has given rise to much fruitful controversy since it was first published in 1938, is

> R. K. Merton, *Science Technology and Society in 17th Century England*. New York: Howard Fertig, 1970

The rise of German academic science in the nineteenth century is described by

> J. Ben-David, *The Scientist's Role in Society*. Englewood Cliffs, N.J.: Prentice-Hall, 1971

The state of science in England in the nineteenth century is described by

> D. S. L. Cardwell, *The Organization of Science in England*. London: Heinemann, 1957

The development of the industrial mode of research is not documented succinctly, but has to be studied in a variety of works describing the growth of particular governmental or industrial research organizations. An overall history of governmental science in the United States to 1940, giving some idea of the policy background, is

> A. H. Dupree, *Science in the Federal Government*. Cambridge, Mass: Harvard University Press, 1957

11

Collectivized science

'We've got no money, so we've got to think.' *Ernest Rutherford.*

11.1 Societal demand

Even though the differences between 'science' and 'technology' (§9.7) between 'research' and 'development' (§10.1) and between 'pure science' and 'applied science' (§10.7) have never been easy to define in principle, an institutional distinction between the academic and industrial modes of research was maintained in practice throughout the first half of the twentieth century. In the past few decades, however, this gap has been steadily closing. Some metascientific observers follow Jerome Ravetz in describing this process as the *industrialization* of science, implying that the industrial mode of research has become dominant. The evidence is, however, that a more general transformation is taking place, to a new *collectivized* form in which characteristics of both the academic and industrial modes are intermingled.

This transformation is often supposed to have come about solely by societal forces acting on science from the 'outside'; as we shall see (§11.2) it is also a natural result of its own internal development. The external influences are obvious. The demand for more and more R & D to meet societal needs (§9.1) has not only had the effect of expanding industrial science on a very large scale: it has also had an immense effect on the scale and style of academic science. As we have seen (§10.5), academic science was never really cut off from society, and its instrumental potentialities were dramatically demonstrated during the Second World War. Academic scientists such as physicists and biochemists turned out to be just as useful as engineers and physicians in dealing with such practical problems as detecting aircraft and submarines by radar, or treating wounds and preventing disease with penicillin and DDT. Above all, the notable success of the nuclear weapons programme suggested a deliberate public policy of fostering and subsidizing 'pure' science without regard for its evident applications.

This was not an entirely novel idea. There was a long tradition of State support for particularly expensive or notable research projects (§14.1). A certain amount of time on pure research was normally allowed to scientists working in many government laboratories, such as the National Physical Laboratory in London and the National Bureau of Standards in Washington, D.C. A similar policy had been followed for many years within the research laboratories of some of the very large firms in the chemical and electrical industries, who even employed a few outstanding researchers, such as Irving Langmuir at the General Electric Company of Schenectady, New York, to work entirely in the academic mode. It became the accepted wisdom that every R & D organization should contain its own nucleus of pure research.

But the policy advocated in the United States in 1945 by Dr Vannevar Bush in his famous report *Science: The Endless Frontier* went much further. This report made a strong case for confirming the wartime government practice of supporting research carried out in the universities by scientists in normal academic employment, rather than running separate, full-time research organizations. In due course, this led in the United States to the creation of new agencies, such as the National Science Foundation, through which large sums of government money could be channelled into academic research. Some of this was for 'pure' science, such as higher mathematics or astrophysics, for which no applications could be envisaged. The subsidies were more lavish for any science whose eventual usefulness could be more plausibly argued, such as biomedical research funded by the National Institutes for Health, or theoretical fluid mechanics supported by the Department of Defense.

The American example was quickly followed by most other industrially advanced countries. The detailed history of a structural transformation like this, spread over several decades, is only of interest to experts in public administration, but the key event in Britain was probably the Trend Report, which appeared in 1962. This brought into being the 'dual support system', by which all academic science is essentially funded by the government, but part of the money is channelled through the University Grants Committee to the universities as whole institutions, and part goes via various research councils to support the research work of individual scientists in particular university departments. This administrative procedure, which had long been used for financing agricultural and medical research, thus brought most academic science under direct government patronage.

The rationale of government support for 'pure' science will be considered when we come to study 'science policy' in chapter 14. All that needs to be said now is that utilitarian motives have always been paramount in the evolution of the present system over the past thirty years. Government spending has always been heavily weighted towards the more 'applied' end of the R & D spectrum (§12.2). For all the fine talk about 'helping you chaps who are pushing back the frontiers of knowledge', government authorities in many countries were persuaded that support for academic research would eventually bring its rewards here on Earth. These benefits would come in the form of some spectacular new invention, such as nuclear power, or would at least 'trickle down' through the technological sphere into the national economy (§13.2). Needless to say, the academics had no difficulty in pointing to numerous cases where this had happened in the past (§9.2), and would even 'promise' to provide similar benefits for the future.

In the last few years, this rhetoric has lost its gloss, and politicians everywhere are imposing much more strictly utilitarian conditions on the science they agree to fund. The demand is for research that will give immediate and definite social, economic, or military results. It is instructive to observe this same policy trend in the Soviet Union, where all science is supposedly undertaken according to a national plan. Nevertheless, even there, the traditional academic mode of pure science has flourished successfully in a number of quasi-autonomous institutions under the effective authority of a highly stratified scientific community (§5.5) and the government is always making efforts to direct pure research towards more useful objectives and to get its results into technological application. Thus, every branch of science is being brought under *collective* control.

11.2 Apparatus

The collectivization of science is not driven solely by societal forces: the same tendency arises within academic science in order to meet the steadily increasing cost of research *apparatus*. Until the present century, this has not been a major constraint in the advancement of knowledge. In an age when research was mostly done by amateurs (§10.3), the most expensive resources they needed were time and energy to spare from the normal demands of making a living. The enthusiastic physician, clergyman, schoolmaster or country gentleman could usually afford to set aside a room in his house as a laboratory or study, to buy a microscope or special chemicals and even to employ a technician or assistant. Scientific apparatus was often skilfully designed, and expertly made, but it was not usually very costly by comparison with the incomes of those who used it.

Some disciplines, of course, such as observational astronomy, were always so expensive that they had to be funded by the State. The financial support given to Tycho Brahe, first by the King of Denmark and later by the Emperor Rudolph II, is well known; this was typical of the patronage of astronomy by the rulers of most civilized nations, including China and India, from the sixteenth century onwards. But William Herschel, the greatest telescope builder and observer of the early nineteenth century, impoverished himself to maintain his expensive hobby until his discoveries made him famous. In many branches of science, it was not the material apparatus that needed external funding, but the long labour of reducing observational material to order, and getting it published in expensive books that might have little public sale.

The provision of special *buildings* for the work of research was seldom seen as a practical necessity until the rise of academic science in the nineteenth century (§10.4). The cost of such buildings then became a major item in the funding of science, although many universities acquired purpose-built teaching laboratories, along with other educational facilities such as libraries and lecture rooms, where scientific staff and graduate students could also do their research. The endowment of a distinctive research laboratory, such as the Cavendish Laboratory at Cambridge in 1870, or the Institut Pasteur in Paris in 1886, was more than a practical facility: it was also a significant symbolic event in the history of science. It brought together a group of scientists with common interests, encouraged them to cooperate in their research, and publicly recognized their discipline as a permanent social institution.

By the 1930s, almost all scientific research was being carried out in large buildings, housing dozens or hundreds of researchers, and fitted out with special services such as piped gases, low-voltage electric power, workshops, animal houses, etc. Facilities such as these were obviously beyond the pocket of most private persons, and could only be provided on an institutional basis to the staff of a university, government department or industrial firm. The cost of individual items of apparatus was also beginning to rise rapidly, often grossly exceeding the funds available from endowments or other regular sources of income. For example, the very first machines for producing beams of high-energy particles, the Cockcroft–Walton linear accelerator at Cambridge and the Lawrence cyclotron at Berkeley, could only be financed from extrabudgetary sources – just as their larger and larger successors have been ever since.

The turning point was the Second World War. Academic scientists moved into an environment where technological devices such as aircraft and telecommunications systems were part of the 'apparatus' of research itself. Every project was of the utmost urgency, and cost could be no excuse for delay. As in industrial research with potentially profitable applications, expenditure on apparatus was easily justified. In the development of radar, nuclear weapons, antibiotics, etc., almost any equipment or facilities that could reasonably be used to speed up the research would be made available. The penury and parsimony that were characteristic of academic science before the war seemed quite inappropriate and outmoded.

At the end of the War, these scientists returned to the universities and immediately felt the need for material facilities on a new scale. Even the surplus equipment from wartime research and operations, such as radar sets and electronic instruments, was a considerable boon, becoming the basis for several new scientific developments such as nuclear magnetic resonance and radioastronomy. To meet the cost of more sophisticated apparatus, and also the technical staff, buildings, administrative services and other facilities to which they had become accustomed, academic scientists called for much more substantial funds than they had received in the past; science not only took up again its regular exponential growth in numbers of scientists, publications and students (§10.2): the amount of money that needed to be spent to keep each academic scientist going at competent research also began to grow rapidly. In some laboratories, in some branches of the physical sciences, the salaries of the scientific staff were soon greatly exceeded by other items in the annual budget for research.

Where could these funds come from? They could certainly not come from the personal incomes of the researchers themselves, and could only be met with great difficulty from the corporate resources, such as endowments, of academic institutions. Private donors and charitable foundations which had supported pure science in the past could not now provide on the scale that was needed. The logic of the situation was unanswerable. Research in the academic mode, with up-to-date apparatus, could only continue if it was supported to a very large extent by private industry or by the state (§14.1). In other words, the *external* forces driving science towards a more collectivized structure were reinforced by the *internal* demand for more and more expensive apparatus and other material facilities.

11.3 Sophistication and aggregation

The most conspicuous characteristics of collectivized science are the increasing *aggregation* of research facilities and increasing *collaboration* in the work of research. These characteristics are not obvious in certain fields, such as pure mathematics, where research is still carried out by individual scientists apparently relating to one another in the traditional academic mode. But there may still be some degree of centralization, in that each researcher may need to justify his or her research programme to some committee of 'peers' (§14.4) to get access to a major instrument such as a computer, or to be given authority over graduate students and postdoctoral assistants. Even when there is no direct coordination of the day-to-day running of research, there may be a gentle but persistent pressure to undertake projects whose outcome might turn out to be vaguely applicable or which are not wildly inconsistent with the current paradigm of the field (§7.3).

In most branches of modern science the increasing *sophistication* of apparatus is an irresistible force. In recent times, and especially since the war, the production of scientific *instruments* has become a major industry. This industry is notably progressive technologically. A modern scientific instrument may have originated as a unique piece of apparatus, personally designed and hand-made for an original observation or experiment. If this is successful, the same apparatus may be copied by other researchers, refined in performance, and eventually marketed commercially.

An 'experimental device' such as a mass spectrometer or electron microscope is thus transformed into a 'scientific instrument', with applications in a very wide range of scientific disciplines.

The transition from hand-made to off-the-shelf apparatus in the practice of research is of considerable epistemological and sociological significance (§2.6). Our main point here is that this transition cannot be avoided. Technologically advanced instrumentation is indispensable in advanced science: there is no returning to a golden age of 'sealing wax and string'. The norm of originality (§6.2) pushes every research worker to the edge of the current 'state of the art' which is most obviously defined in terms of the latest, most powerful instruments in the field. Competition between researchers fuels commercial competition between instrument manufacturers, who are under continuous pressure to improve their products, almost regardless of price. It is true that the latest model of a standard laboratory instrument such as a gas chromatograph may cost twice as much as the equivalent model of 5 years ago: but it may be ten times as precise, reliable and speedy in use. Not to have such an instrument as soon as it becomes available is to risk serious loss of competitive vigour in the race to a scientific discovery.

Scientific and technological progress is also generating a much greater variety of precision instruments that can be applied to the solution of any particular scientific problem. For example, the traditional technique of identifying a chemical compound by optical spectroscopy can now be complemented by infra-red spectrometry, nuclear magnetic resonance spectrometry, mass spectrometry, etc., each providing distinctive and significant information. It would be absurdly costly to give every research worker a whole panoply of these expensive instruments for his or her personal use. They can only be made available if they are shared over the research activities of a relatively large organization. The *aggregation* of research facilities is a characteristic tendency of all advanced science, whether it is academic microbiology or industrial solid-state physics.

11.4 Collaboration

In the extreme of sophistication, the apparatus of research becomes too large and complicated to be manipulated by a single individual, or the variety of instruments to be used in a particular project calls for a wider range of intellectual and practical expertise than any single research scientist can master. This means that research projects can only be undertaken by the active *collaboration* of a number of researchers.

There is nothing new, of course, in the employment of technical assistants and technicians to do the practical work of research. In some cases, the assistant has even surpassed his master: Kepler was employed by Tycho Brahe to make observations and calculate astronomical tables, and thus had at hand the means to discover the laws of motion of the planets. By tradition, graduate students and post-doctoral assistants have had to learn the craft of research by apprenticeship to an established scientist (§10.4) before gaining an independent academic post. In general, however, these relationships were distinctly unequal, whether in education, social class, experience or age, so that the master researcher would usually treat his assistants as subordinates whose actions were little more than part of the mechanism of the research apparatus, rather than as scientific colleagues.

The modern phenomenon of *team research* is new in that it involves direct collaboration between scientists of relatively equal standing. In an interdisciplinary project, high-level specialists from several different academic disciplines (§5.3) may have to work closely together. Thus, an investigation of the underground disposal of radioactive wastes would call for expertise in nuclear physics, materials science, geology and hydrology, not to mention medical, economic and political considerations.

Team work is, of course, characteristic of all technological development (§13.3), but it is not confined to 'applied' research in the industrial mode. A striking feature of contemporary science is the development of enormous instruments, such as particle accelerators, space probes and radiotelescopes, devoted entirely to 'pure' research. These *big science* facilities are not only immensely costly to construct and run; they can only be put to good use by the cooperative efforts of large numbers of fully qualified scientists taking on a variety of narrowly specialized rôles within a single project. Thus a single experiment in high-energy physics is usually the collective enterprise of as many as a hundred scientists of established academic standing (§10.4) dividing between them the labour of designing and building the detectors, writing computer programmes, aligning the beam, monitoring experimental runs, interpreting the data, and so on. And in the end, the results of this experiment are published as just one primary paper (§4.1), with a hundred co-authors each seeking some degree of personal 'recognition' (§5.1) for their contribution to knowledge.

11.5 *The collectivization of science*

Science is thus being 'collectivized' in two different senses of the word. On the one hand, almost all research nowadays is being carried out within the framework of quite large organizations, such as research laboratories, institutes, or university departments where the facilities for research are shared by dozens or hundreds of scientists. In many cases, research projects are undertaken by groups or teams of researchers who have limited control over the resources they use, and cannot claim personal responsibility for what is attempted or what is achieved. In other words, the extreme *individualism* embodied in the academic ethos (§6.3), and the norms associated with it, is no longer consistent with the realities of the scientific life, where *collective* action is now the rule. The internal sociology of academic science is being transformed (§12.5). The conventional description of the scientific community as a republic or oligarchy of autonomous scientists, exchanging communications for personal recognition (§5.2) is not yet out of date, but it must be radically modified to take full account of the structures that have grown up to coordinate and *manage* scientific work (§12.4), even in its 'purest' and most academic mode.

On the other hand, the incorporation of academic science into an expanding 'R & D' system, drawing funds from the central government and private industry, represents an equally profound transformation of the external sociology of science. Academic science is losing its place as an autonomous social segment with its own standards and goals (§10.5), and is being brought under 'collective' control. Instead of being treated as an independent source of unpredictable social influences, it has come to be regarded as an instrument of deliberate societal action. Science has thus moved from the periphery of society towards the centres of power, and is apparently becoming an organ of the state apparatus, the ruling class, the military–industrial complex, or whatever it is that governs our lives in an advanced industrial country (§14.6).

Once again, the conventional description of academic science as a distinct social institution is not entirely obsolete, but needs to be considerably revised to take account of the effects of external influences on the objectives and performance of research. These influences do not arise solely from direct societal demand. To get the expensive apparatus they need for their research, the leaders of academic science have had to give an account of their activities in quasi-economic language and justify their projects in utilitarian terms. These diverse consequences of the general process of collectivization will be the main theme of the next three chapters.

Further reading for chapter 11

A straightforward account of the changes in the societal rôle of science in recent years is given by

H. Rose & S. Rose, *Science and Society*. London: Allen Lane, 1969

The 'industrialization' of science, and its consequences, is discussed by

J. R. Ravetz, *Scientific Knowledge and its Social Problems*. Oxford: Clarendon Press, 1971 (pp. 31–68)

A racy account of the pressures of 'big science' funding is given by

D. S. Greenberg, *The Politics of American Science*. Harmondsworth: Penguin Books, 1969 (pp. 81–165)

The famous report 'Science the Endless Frontier' by Dr Vannevar Bush is reprinted in

W. R. Nelson (ed.), *The Politics of Science*. New York: Oxford University Press, 1968 (pp. 26–55)

12

R & D organizations

'In modern science the era of the primitive church is passing, and the era of the Bishop is upon us. Indeed the heads of great laboratories are very much like Bishops, with their association with the powerful in all walks of life, and the dangers they incur of the carnal sins of pride and lust for power.'

John von Neumann

12.1 Science as an instrument of policy

As we have already remarked (§9.1), to the eyes of the general public, science is simply one of the components of 'science and technology', which is primarily an instrument in the hands of society. This instrument can be used to do whatever society wants, over a very wide range. These wants are impossible to list in full, and have very varied sources of motivation, such as:

Meeting basic human needs, in the form of food, shelter and health
Making war, or otherwise serving the purposes of the nation-state
Making profits for competitive industry, through technological innovation
Improving the quality of life, by eliminating human drudgery and environmental pollution
Solving social problems, such as overpopulation and economic underdevelopment.

In the past half century, the vague Victorian belief in science as a source of 'progress' has been transformed into an established doctrine. It is now widely held that distinct advances towards political, military, economic and commercial ends can best be achieved by deliberately fostering the right sort of research and development (§11.1). This fostering is to be done primarily by providing financial resources, either directly from the State (§14.1) or from large-scale industrial corporations, and channelling them to the appropriate R & D organizations through administrative bodies such as research councils, universities, government departments or the research divisions of private firms. Although it is accepted, in principle, that the outcome of R & D cannot be guaranteed, the general opinion is that this is an investment with a very favourable average return (§13.2), especially if it is efficiently directed towards the most salient needs of the moment. Thus (according to this doctrine), it is essential to have a good *science policy* (§14.2), to ascertain these needs, determine their relative priorities, and set up a suitable administrative machine to feed resources to the corresponding sector of the R & D system of the nation.

From this point of view, the only important factors in the social relations of science are economic and political. Research is taken to be a practical enterprise to be ventured upon in the same spirit as any normal commercial, political or military operation, and assessed by the degree to which it achieves the objectives for which it was undertaken. The internal structure of R & D organizations seems unproblematic or purely sociotechnical, like the internal structure of an army or bank. The use of 'science' is taken to mean no more than the application of 'scientific method', 'scientific knowledge' and 'scientific techniques' to whatever problems are to be solved. Scientists are considered as no more than technical workers (§15.4), trained to be experts in a variety of highly specialized tasks, and employed by R & D organizations to make discoveries and inventions. The main problem for society is thus to ensure that there is an adequate supply of suitably qualified people, and that they are efficaciously deployed for the work to be done: just how they actually do this work is their own business, provided that they do not question its ultimate objectives.

The instrumental function of science has always been considered the basic principle of its societal rôle. Collectivized R & D organizations are what you get if you set out systematically to implement this principle in *practice*. This general notion of how to put science to use as an instrument of policy is the commonplace opinion throughout public life in all economically developed countries. This is the way that politicians talk about science to government officials, company executives talk to their accountants, industrialists talk to generals and admirals, and newspapers address their readers in editorials. Although science is really a much more complex social institution, internally and in its external relations, than these notions allow, it is bound to become, in the long run, what people think it is or ought to be.

12.2 The spectrum of relevance

As we saw in chapter 10 the traditional sharp distinction between 'pure' and 'applied' science is no longer valid. Instead of falling into two clearly defined categories, corresponding to the 'academic' and 'industrial' modes of research, modern R & D organizations range along a continuous spectrum of 'relevance'. The sub-divisions of this spectrum are variously termed, but the following nomenclature is typical.

Basic science is knowledge-oriented, and is undertaken as if 'for its own sake'. Its objectives are expressed in phrases such as to 'uncover fundamental principles', to 'explore nature', or to 'understand how things work'. In other words, this is academic science at its 'purest', without any utilitarian motives. Any connections that it may be supposed to have with present or future technologies are merely conjectural.

Strategic research is also knowledge-oriented but it has a conscious utilitarian purpose, such as the determination of the facts and theoretical principles relevant to existing or projected technologies. Even though it may not be directed towards the solution of any specific practical problem, it is expected to prove its value in the long run by its contribution to practice. A prime example would be the detailed investigation of the physics of plasmas, in the hope of applying this knowledge eventually in the use of nuclear fusion for energy generation.

Mission-oriented or *targetted* research covers investigations that are much closer to the actualities of current practice or current problems. A typical example might be a detailed study of the relation between exposure to radiation and cancer, or research on chemical reactions in the upper atmosphere to determine the effect of fluorocarbons on the ozone layer. By definition, this sort of research has a clear utilitarian motive, and is expected to produce results that can be put into action in a practical way.

Technological development is the final category, and covers all the work of technical improvement, detailed design and testing of components, trials of prototypes, etc., that is needed to bring a new product or process into regular use. This end of the R & D spectrum, where immediate utility is the paramount consideration, used to be the sphere of the inventor, engineer, or clinical innovator, but now employs many people trained primarily as research scientists.

These sub-divisions are not precisely defined or mutually exclusive. It is often very difficult to place a particular R & D organization, or even a particular R & D project, into one or another category. In practice, institutions, research groups, professional societies and scientists themselves straddle the sub-divisions, or move from one part of the spectrum to another, from year to year, or even from day to day. A major research organization, such as the Atomic Energy Research Establishment at Harwell, may be undertaking R & D projects covering the whole range of relevance, from the most basic theoretical physics to the engineering development of an installation for desalting seawater.

It is not safe to assume that the spectrum of relevance represents a gradient from 'academic' to 'industrial' R & D organizations. Many large industrial firms, such as chemical manufacturers, undertake a great deal of strategic research and may even support small, in-house, basic science units; many universities, on the other hand, are deeply and very urgently involved in the commercial development of various forms of biotechnology. Nor is 'relevance' a reliable indicator of the degree to which a branch of science is collectivized. Elementary particle physics is a perfectly basic science, of no known utility, yet it is highly organized into research teams; the development of technical innovations in clinical medicine is usually highly individualistic, despite its obvious pragmatic goals.

Nevertheless, the position of an R & D organisation along this spectrum is a rough indication of its closeness to the needs and values of everyday life. The notion of 'relevance' cannot be strictly defined or measured, but is always invoked in discussions of the social function of science, and is often the deciding factor in science policy.

12.3 The philosophy and methods of R & D

Despite the wide range of social relevance of the problems they tackle, R & D organizations share a common philosophy. This philosophy can best be described as the traditional 'method' of academic science, as set out in chapters 2 and 3. Observation (§2.2), experimentation (§2.8), the collection of empirical data (§3.2), theorizing (§2.3), prediction (§3.6), and corroboration (§3.7) all play their part. Rational analysis on the basis of established knowledge (§3.8) is held to be the ideal, but the significance of serendipitous discoveries (§2.5) and revolutionary paradigm shifts (§7.3) is clearly recognized by all experienced R & D workers.

But R & D cannot be clearly separated from technological practice at the more relevant end of the spectrum. It thus tends to come into conflict with rather different philosophies, such as the aesthetic considerations of engineering design, or the ethical considerations of clinical medicine. One of the major characteristics of R & D is its insistence on its 'scientificity' – that is, its attachment to the regulative principles of scientific work (§3.10), and all that follows from them. Whether or not this is justified in narrow epistemological terms, it is the source of much controversy over the relationship between scientific and other human values (§16.5). Thus, for example, a clinical trial that is scientifically justified, to test the efficacy of a new drug, may not be permitted on ethical grounds because it seems to put certain patients at risk. Issues such as these put the philosophy of science to far more serious test than the eternal debate on whether science describes things as they 'really' are (§3.9).

The 'scientificity' of R & D (§10.1) is emphasized by the way that it draws upon the conceptual content of the basic sciences right across the whole range of relevance. An engineer developing a rocket motor knows that what he does must be consistent with the laws of thermodynamics, even though he may have great difficulty in putting this knowledge to use. A cancer specialist is bound to interpret his observations according to the principles of molecular biology, genetics, virology, etc., and would be most reluctant to propose hypotheses that did not use the concepts of these sciences. Each branch of R & D work tends to develop its own body of specialized lore (§3.3), but there is no boundary in principle to the *scientific* knowledge that might be thought applicable to a particular type of problem.

It is noteworthy that the day-to-day techniques, instrumentation (§2.6, §11.2), theoretical concepts, educational background and other resources of R & D workers are more or less uniform at the laboratory or research group level, right across the range of relevance. The main difference is that as one approaches the more 'applied' end of the range, one usually finds greater diversity of *interdisciplinary* approaches. Thus, for example, the scientists studying the basic physics of semiconductors in a university physics department will probably all have been trained in physics, whereas a team brought together to develop a new semiconductor device in the research laboratory of an electronics firm might include chemists, electrical engineers and even mathematicians. But this may be simply a consequence of the way that scientific disciplines are defined academically, and does not apply when new basic disciplines such as molecular biology are coming into existence (§7.2). According to the *Starnberg hypothesis* (named after the German university where it originated) scientific disciplines grow out of basic research, but as they reach maturity, and come under the sway of a well-founded paradigm, they are *finalized* – that is they are directed towards practical ends, and become the basis of new technologies (§9.6). There are historical cases that fit this pattern of change within the R & D system, but the argument as a whole is more controversial than it sounds because of its connections with other concepts in the sociology of knowledge (chapter 8). It may also be misleading to represent technological development as the application of well-founded scientific knowledge with a firmly established theoretical paradigm: as R & D becomes more relevant to action, it tends to become more *pragmatic*, and to assess the validity of knowledge simply by what happens to work in practice.

12.4 The management of R & D

Although R & D is still based philosophically on the traditional 'method' of academic science, it no longer conforms internally to the sociological structure in which this 'method' originally developed. The collectivization of science is producing a transformation of this structure comparable to the general historical transformation from *gemeinschaft* to *gesellschaft* – from 'community' to 'corporation' – described by Max Weber.

It is true that many scientists still hold permanent, intellectually independent posts in universities, and win scholarly preferment by their public contributions to knowledge (§10.4). The communication system of science (chapter 4) proliferates with publications, and the competition for recognition and authority (chapter 5) is said to be fiercer than ever. The institutions and ethos of the academic mode of science are still very much alive, especially in those branches of basic research, such as pure mathematics, which seem pretty useless and make no financial demands for expensive apparatus (§11.3).

But these organizational forms are being modified, even in universities and other institutions devoted primarily to basic and strategic research, towards the structures that used to be found only in industrial research organizations (§10.6). Most large R & D organizations now find that their work and their workers must be *managed*, according to rational administrative principles (§15.4), rather than simply being fostered, encouraged, or permitted to evolve in response to individual initiatives. The craft of *research management* has thus developed into a specialized branch of the crafts of industrial management and public administration, with a professional literature all of its own.

The sociology of R & D is now treated mainly as if it were a specialized topic in the sociology of organizations. It must not be assumed, however, that the application of the general principles of management – however rational or humane – will automatically improve the efficacy, productivity, or accountability of R & D organizations. Scientific research remains a social activity of a peculiar kind, with its own characteristic styles of leadership, intermediate objectives, institutional loyalties, personal incentives and vocational commitments.

There is no reason to suppose, for example, that a very large R & D organization must be a hierarchical *bureaucracy*, run on the lines of an army, a civil service or a large business. The material and informational necessities of scientific work are quite different from the necessities of military, administrative or commercial operations, and call for quite different relationships of authority and responsibility. Indeed, there are few R & D organizations that are really large by the standards of modern industrial society, and most of them could be managed quite satisfactorily by relatively informal methods, without recourse to strict bureaucratic procedures. This is evident from direct studies of typical R & D organizations, such as government establishments and industrial laboratories. The actual power structure within such an organization may range from the centralized autocracy of an omnipotent director, through loosely knit federations run by oligarchies of group leaders, to an almost anarchical democracy in which all established scientific workers have a say in what goes on. Such structural variations may not seem important to the outsider wanting to use the organization as an instrument of policy, but they are obviously very significant for the researchers themselves and for the quality and/or relevance of the results they produce.

12.5 The internal sociology of collectivized science

The internal sociology of organized R & D deviates markedly from the ideals of academic science. This becomes quite evident if we look at the Mertonian norms

(§6.2) and see how well they are actually observed in modern R & D. In many respects, the behaviour dictated by these norms is clearly inconsistent with the interests and structural principles of collectivized scientific institutions.

The norm of *communalism*, for example, is in direct conflict with the proprietorial interests that motivate almost all R & D at the development end of the spectrum of relevance. The work is done precisely to acquire knowledge that is not available to enemies or competitors. The obligation on the scientist to make public all that he or she discovers is therefore severely restricted. But restrictions demanded by military or commercial secrecy are not imposed solely on 'applied' science. Many industrial and defence laboratories do a great deal of strategic and basic research which only gets published if it seems to have little utility from their point of view. The constraints on communalism in R & D are very pervasive. It is not sufficient to say that university scientists working on industrial or government grants are 'normally permitted' to publish their findings; the very fact that permission to publish has to be obtained from an external authority shows that this norm is no longer operational for these scientists.

Again, the norm of *universalism* is not consistent with the hierarchical structure of authority in a managerial organization. The manager is given the power to insist that certain projects should be undertaken, that certain resources should be assigned to a proposed investigation – and, in the end, that certain research results should be regarded as fully established, regardless of the opinion of any scientific subordinate. It has always been difficult, even in the most liberal era of academic science, to give a sufficient hearing to scientific voices from outside the upper strata of specialists in a particular field (§5.5). In collectivized science the traditional deference to the 'charisma' of an eminent scientist may be reinforced by official power. The senior professor or academician who is also the director of a large institute – this is the typical situation in the Soviet Union – is thus in a position to impose his or her intellectual doctrines on all subordinates.

The norm of *disinterestedness* was never easy to obey, even in 'pure' research undertaken 'for its own sake'. This norm is obviously inapplicable in technological development, and comes under heavy pressure in any R & D organization which is specifically oriented to the success of a particular mission or the furtherance of a particular branch of science. Thus, for example, scientists employed by an atomic energy agency on strategic research relevant to the safety of nuclear power would not present their research results in the same way as they would if they were working on behalf of an environmental protection group (§15.5). A public stance of humility and objectivity becomes more and more suspect against the background of paid employment and team loyalty (§11.4), which is the reality of all collectivized scientific organizations, however devoted they may be to the pursuit of knowledge 'for its own sake'.

12.5 The internal sociology of collectivized science

The novelty of research lies as much in its intentions as in its results. The norm of *originality* should apply as much to the proposal to undertake a particular investigation as in actually carrying it out. In a bureaucratic R & D organization, personal autonomy in the choice of research problems and methods, which is one of the essential features of academic science (§10.5), is subject to higher authority. The research manager has the right to approve or veto research projects, and hence to limit the exercise of scientific originality in more junior positions. This is such a serious matter in basic research that a complicated machinery of peer review of research grant proposals has been set up, in order to take these decisions collegiately (§14.4). But from the point of view of the individual scientist, the collective nature of such decisions is little consolation. It is not easy to demonstrate one's originality as a scientist if one's research plans are subject to the authority of other people, however knowledgeable and enlightened they may be. It is also difficult to show one's imaginative powers if one is a member of a large team working on a single project (§11.4) or if the research one is doing has been commissioned by some external agency.

Finally, there is the norm of *scepticism*, which enjoins scientists to take nothing on trust. But this norm often conflicts with the institutional loyalties and 'house doctrines' that tend to establish themselves in formal social institutions. There are cases where scientists have discovered (or believe they have discovered) grave weaknesses in the knowledge base underlying the practical policies of a large organization, and then have had great difficulty in getting their sceptical opinions heard. The problem of protecting the personal rights of the 'whistle-blower' in a mission-oriented R & D organization (§15.6) is clearly linked with the problem of observing the norm of intellectual scepticism in all collectivized scientific work.

The Mertonian scheme of norms was never, of course, more than an idealization of the traditional rules of academic science. The ethos that it defines (§6.3) had many of the characteristics of a self-serving ideology (§6.4, §10.7). Very few academic scientists could ever live up to this ideal in practice. Nevertheless, it was not an utterly impossible standard to aim at, and could be referred to as the basis of many of the actual rules and conventions of scientific life. In recent years, the validity of these norms has been severely questioned by many sociologists of science, who point to the incidence of secrecy, excessive authority, material interest and conformity in the scientific life of today. Was science always like this, or has it, as I am here arguing, gone through a substantial transformation in the past few decades? Whichever way we look at it the modern collectivized R & D system does not conform closely to the ethos of academic science, and is not structured internally to foster that ethos (cf. §15.4).

Further reading for chapter 12

Much modern writing about 'science' actually describes the way of life of R & D organizations.

The general characteristics of the contemporary R & D system are described by

    J. J. Salomon, *Science and Politics*. London: Macmillan, 1973 (pp. 71–115)

    A. Weinberg, *Reflections on Big Science*. Oxford: Pergamon, 1967 (pp. 123–74)

    M. Blisset, *Politics in Science*. Boston: Little Brown & Co., 1972 (pp. 162–96)

    L. Sklair, *Organized Knowledge*. London: Hart-Davis, MacGibbon, 1973 (pp. 13–100)

The conflict between the 'academic' and 'industrial' norms is a regular theme in books on the management of industrial research, such as

    S. Marcson, *The Scientist in American Industry*. Princeton, N.J.: Princeton University Press, 1960

The diversity of managerial structures in R & D organizations is indicated by

    T. Shinn, 'Scientific Disciplines and Organizational Specificity: the Social and Cognitive Configuration of Laboratory Activities', in *Scientific Establishments and Hierarchies*, ed. N. Elias, H. Martins & R. Whitley, pp. 239–64. Dordrecht: D. Reidel, 1982

The problem of personal autonomy in R & D is discussed by

    J. M. Ziman, 'What are the options? Social determinants of personal research plans.' *Minerva*, **19**, 1–42 (1981)

13

The economics of research

'The story is told that Sir Robert Peel, the Prime Minister, visited Faraday in the laboratory of the Royal Institution soon after the invention of the dynamo. Pointing to this odd machine, he inquired of what use it was. Faraday is said to have replied "I know not, but I wager that one day your government will tax it".'
          L. Pearce Williams (in *Michael Faraday*, London: Chapman & Hall, 1965)

13.1 Costing the benefits

The purpose of R & D is to provide benefits. But how should the value of these benefits be assessed? It is all very well to say that research on insecticides has resulted in improved crops of bananas, but was the improvement worth the cost of the research? Research costs real money and its outcome is very uncertain. It might have been more profitable to invest the money in a new plantation. Without some rough estimate of the relative balance of costs and benefits, the use of science as an instrument of policy (§12.1) is based solely on blind faith.

The *inputs* to R & D can easily be quantified in money terms. The cost of employing researchers and providing them with suitable apparatus, buildings, technical staff, telephones, travel to conferences, and so on would normally appear as line items in the financial accounts of the corporation or agency supporting the laboratory. The difficulty comes when one tries to measure the *output* of an R & D organization in the same terms. What is the value of a scientific paper in pounds, or dollars, or roubles?

The *economics* of R & D is normally treated as a specialized sub-discipline, outside the domain of academic metascience: yet it is really an essential dimension in the external sociology of science and technology. In a society where financial considerations – e.g. profits – are the major motivation for social action and cultural change, one must ask how these considerations enter into policy for R & D. In particular how does this type of activity figure in the calculations of capitalist enterprises in a market economy?

This is not an easy subject to grasp comprehensively because of the enormous range of benefits generated by science. A large proportion – as much as 40% according to some estimates – of expenditure on organized R & D goes for the invention or improvement of weapons of war. This item may look good when set against the commercial profits of manufacturing and selling arms, at home or abroad, but must surely be counted as producing social *disbenefits* in a larger analysis. On the other hand, the value of good health is incalculable, so that almost all medical research seems eminently justifiable in economic terms. The tendency has been, therefore, to concentrate on a rather limited set of problems, where more precise accountancy is possible, even though these are not fully representative of what goes into, or comes out of, R & D in a modern society.

13.2 Macroeconomics of R & D

Considered overall, research undoubtedly pays handsomely. By any reasonable measure of 'wealth', advanced industrial societies have greatly enriched themselves by deliberately fostering and exploiting scientific knowledge and technique. Any other conclusion from the history of science and technology seems perverse. For example, whole industries, such as electrical and electronic engineering, which now account for a significant fraction of the total economic activity of the nation, derive originally from basic scientific discoveries (§9.2).

Nevertheless it is not at all easy to set a figure to the value of these benefits, even in quite specific cases. The fundamental difficulty is in assessing the comparative values of goods that have been substantially changed in design, use, efficacy and market penetration after the application of the results of research. To give an extreme example: 50 years ago my grandfather, a consultant physician, might have prescribed those famous pink pills, 'worth a guinea a box', knowing that they could merely alleviate a serious condition: nowadays, a course of antibiotics, costing a few pounds in present-day money, would be sure to cure the disease completely. Even allowing for inflation, the historic costs and such-like accountancy entries for these nominally comparable goods bear no relation to their real values.

The standard case study of the profitability of R & D deals with the development of hybrid corn, a commodity whose intrinsic value for food and fodder can be regarded as reasonably constant over a long period of years. In fact, it took about 25 years of strategic and mission-oriented research by various American agricultural research institutions before this potentially productive idea could be developed for widespread use in the 1930s. Since then, however, the substantial expenditure incurred in this long and uncertain project has proved enormously profitable to the American farming community. According to the calculations by Zvi Griliches, this 'investment' in R & D has yielded a return of hundreds of per cent per annum, apparently in perpetuity.

But this is only a notional calculation, since it cannot take into account all the research efforts in other directions that have not generated successful technological innovations. Perhaps one should consider all agricultural research and the total improvement of agricultural productivity as the 'input' and 'output' variables. But then one would have to include a number of other factors, in addition to R & D effort, that contribute to the growth of agricultural production. These might include the investments required to bring new land under cultivation, the displacement of old methods, such as horse-drawn ploughing, by existing, more up-to-date techniques, such as the use of tractors, and the employment of a better educated and more competent labour force. Some economists have tried to estimate the total contribution of all changes of this kind to the growth of the gross national product (GNP) and then argued that the residual term must be due to the technological innovation arising out of scientific R & D. Unfortunately, these calculations have not proved very convincing as a means of assessing the cost/benefit ratio of R & D, and are interesting mainly in demonstrating that the input from science is not the sole, nor even the dominant, factor in economic growth.

Some economists, notably Kondratiev and Joseph Schumpeter, have argued, however, that technological innovation is not merely the main growth factor in the world economy, but is also the driving force of the regular cycle of booms and slumps that has been observed over the past century or so. The theory of *long waves* emphasizes the ambivalent effects of new technologies, which can induce economic depression by the displacement of labour in established industries, as well as providing opportunities for profitable investment in completely new industries. For this mechanism to work, there has to be positive gearing between the general economic cycle, and cycles of inventive activity as measured, for example, by the number of patents taken out in various industries. Although the evidence in support of this theory is much disputed amongst economists, it has lately become a topic of major interest for students of the social relations of science and technology. But the whole issue of the causes and consequences of technological change is much wider than the economics of R & D as such, and would take us too far away from strictly scientific and metascientific questions.

Another line of macroeconomic analysis that had a certain vogue in the 1950s and 1960s was to compare the rate of growth of GNP with national expenditure on R & D, from country to country. Unfortunately, the correlation between these two variables did not prove to be very strong. This was a severe blow to the widely held doctrine that money put into R & D would almost automatically 'trickle down' (§11.1) into technological innovation and thus produce industrial benefits that would eventually show up in the national financial accounts. It seemed reasonable to suppose that countries that spent more on R & D would be enriched in due proportion. The fact that this did not appear in the comparisons between different countries could only be explained by reference to a variety of subsidiary factors such as relative expenditure on military and civil R & D, the balance of effort as between basic research and technological development, market opportunities, the transfer of technology by purchases of patent rights from abroad, and so on, until the original point at issue was obscured beyond useful analysis.

There is thus no general measure of the *profitability* of R & D, aggregated over a whole nation. Science policy cannot be based firmly on a general principle of the form, 'The optimum rate of expenditure on R & D, to get the maximum benefit in economic growth, is about x per cent of the national income'. For many advanced countries, the present value of x happens to be about 2, but there is no reliable evidence that this is indeed an optimum figure. This is a particularly important matter for the less-developed countries, who have to take harsh decisions on whether to invest their hard-won savings (or foreign loans!) in expanding their very modest R & D facilities for possible long-term benefits, or whether they should put their money into more directly productive resources such as factories and roads.

13.3. The sources of invention

Almost all productive and service industries, from steel-making to mass advertising, are now more or less 'scientific' (§§9.2–9.4). They all derive the means for technological change from R & D organizations (§12.1) sited in major firms or in semi-public institutions such as universities and government establishments. But that certainly does not mean that all product or process *innovation* arises directly from organized R & D within the industry that benefits from it. In any calculation of the economics of research, the flow of ideas and information between the various segments of the economy is a major factor.

Detailed case studies of significant technical innovations of recent years have shown, for example, that many of these still originate from the imaginative initiatives of individuals working outside the formal R & D organizations of big business or big government. Important inventive ideas and novel techniques still come from small venture firms, from professional practitioners such as design engineers and clinicians and even from the occasional 'backyard inventor'. The detailed structure of the *patent* system as a means of fostering such efforts is clearly very significant in the economics of R & D, especially since this legal instrument can be used to protect an undeserved monopoly as well as to reward enterprise.

But the original conception of an invention is only the first stage in a long process where large-scale organizational considerations become more and more dominant. To get a brilliant invention such as xerography or float glass into marketable production usually requires an immensely laborious and expensive process of technological development (§§10.1, 12.2) far exceeding the financial resources of the

13.3 The sources of invention

individual inventor or small firm. It is often said, for example, that an idea that cost £1 to discover originally, will need £10 worth of targetted research (§12.2) to get to the stage of a working prototype and the expenditure of £100 to develop this into a salable product. Multiply these figures by a million to get some notion of the cost of applying science in modern industry. This need for large chunks of venture capital to exploit the initial products of research and/or invention is a familiar fact of life for all advanced industrial nations.

Technological development also calls for *intellectual* resources far beyond the competence of any single person (§11.4). The inventiveness of a modern industrial corporation depends in large part on its ability to concentrate factual information, formal principles, tacit skills and imaginative insights from a wide range of scientific disciplines and technical professions on a single problem or project. This knowledge has to be available, within the corporation, and yet it could not all have been generated by any amount of internal effort, nor simply purchased as needed.

To illustrate this point, consider the history of the jet engine, which was developed independently and secretly in Britain and Germany during the Second World War. In each country, jet aircraft came into service after about 5 years of intensive engineering *development*, the testing of prototypes, redesigns, etc., etc., preceded by about 5 years of *mission-oriented research* (§12.2) needed to establish the feasibility of the idea. This was, so to speak, 'in-house knowledge' generated within the R & D organizations on each side. But it was founded upon more than 50 years of *strategic research* on the thermodynamics of internal combustion engines and turbines, and this in turn rests on several centuries of *basic research* on classical physics and continuum mechanics which culminated in the discovery of the concept of energy in the middle of the last century. Thus an apparently distinct and self-contained R & D enterprise really depended upon the whole range of elementary and advanced scientific knowledge which permeates the technical activities of all modern industry (§9.7) – knowledge that goes back to basic, academic scientific research, and to practical technological experience in many countries over long periods of years.

The apparent contradiction between two famous studies – project *Hindsight* and project TRACES – is thus easily resolved. The former study, which seemed to show that the major proportion of military technological innovations came out of engineering development, looked back only a couple of decades: the latter study showed that these same innovations incorporated scientific knowledge derived from academic research going back to much earlier dates. In other words, the intellectual sources of an invention are not solely personal, nor the distinct property of any private corporation or nation; to a large extent they come from the collective efforts of the world's scientific community and are already in the public domain, available as a free good for any practical purpose.

13.4 The microeconomics of research

Now let us consider the logic of the situation for the board of management of a single firm marketing technically advanced products. How much, of what kind of R & D should be undertaken by the firm itself? It is easy to decide what should be done at the two extremes of the spectrum of relevance (§12.2). At one extreme the firm must have its own in-house facilities for the *development* of appropriately chosen innovations from the stage of demonstrated feasibility right through to production for the market. Most of the specific information required for this work may be potentially available as a commercial commodity – either secret, or tied up by patents – but can also be generated internally as an integral part of the usual processes of design, prototype testing and quality control that lead to full-scale manufacturing. It is largely a question for the financial accountants of the company how much should be invested in an all-round capacity for technological development or how much R & D information ought to be bought from outside.

At the other extreme, the *basic* research results that might prove relevant in the development of a novel product are already available as public scientific knowledge, so that there is no profit to be gained from undertaking R & D of this kind, except perhaps as a wild speculation in the hope that something useful might turn up. The same argument would apply to *strategic* research into the scientific foundations of the industry, most of which would also have been openly published long before it could be put to practical use. It is true that some of the results of such research might be indirectly beneficial in improving the technical sophistication of product and process development, which is often done by engineers and other technical practitioners who are no longer in touch with the latest discoveries in 'pure' science, but that might prove an expensive luxury in hard times.

Participation in strategic research in relevant fields does, of course, offer the tempting prospect of making a discovery that will eventually generate a highly profitable innovation. But this prospect must not be too distant: at a discount rate of 5%, the horizon of the accountable future is about 15 years, so that it is very difficult to justify an uncertain investment with a longer time to the pay off. In any case, this sort of knowledge is continually changing, so that it may be wiser to assume that, if there is going to be a major conceptual breakthrough comparable with the discovery of nylon, or the invention of the transistor, up-to-date scientific expertise can be bought into the company by hiring recent graduates, or offering consultancies to well-informed academics. As the disciplinary mix of technological development itself tends to change with time – for example, the physics relevant to electronics has shifted from the vacuum to the solid state – it could be an expensive mistake to build up a large capability for strategic research in a field that may, within a decade or so, turn out to be irrelevant to the main activities of the firm.

To what extent, then, should the firm undertake *mission-oriented research*? It should

13.4 The microeconomics of research

be worthwhile to make a systematic study of the current methods of production, looking for openings for significant improvement and expecting to make a number of minor inventions and innovations. Ideally, such investigations should be open-ended, without preconceived notions on the problems likely to turn up, and with no certainty of a positive return on any particular project. This sort of exploratory research within a closely defined context may not be very glamorous, but it ought to throw up enough novel ideas to pay for itself when these are developed and put to use. Ideas of this kind are seldom made public immediately if they look at all promising, and often turn out to be extremely profitable for the firm that can exploit them competently. Technologies evolve – and eventually transform themselves out of all recognition – by the increment of innumerable modest innovations that are sometimes difficult to detect by the inexpert eye. Any progressive firm must be able to keep up with this evolutionary change by its own efforts in R & D.

Nevertheless, as everybody knows, only a small proportion of promising inventions prove to be successful and profitable, and the process of developing them is usually laborious and costly. A sum of money spent on an activity with a relatively low probability of getting any return at all cannot be called an *investment*: R & D is much more like a *lottery*, where most of the tickets will yield nothing at all, even though a few can win very big prizes. This is illustrated, for example, by the history of the National Research and Development Corporation (now the British Technology Group), a quasi-non-governmental organization to finance the development of inventions: out of the hundreds of projects it has supported, only a few have proved reasonably profitable, but just one – the drug cephalosporin – has paid back all and more of the total expenditure on all the other projects. It may be that NRDC was not efficiently managed; nevertheless a very skew distribution of profitability is characteristic of all R & D enterprises.

From an accountant's point of view, therefore, technological development is an essential cost of staying in business; but open-ended R & D is a very risky undertaking, with no secure return, and very sensitive to imponderable factors such as the hunches of research managers and the impact of new scientific discoveries. In hard times it can look just the sort of unproductive element that can be cut down without immediate loss, even though it might pay off very well if it were steadily pursued over a long period. A major industrial corporation, such as a multinational company or a nationalized industry, can afford to undertake R & D on such a large scale that the risks and returns are spread over many projects and many years of effort. But mission-oriented research on a smaller scale is essentially a speculative enterprise, outside the prudent calculations of the rational man of business. As the current boom in biotechnology shows, such enterprises can get support in the private sector of a market economy, but only in the form of venture capital, from wealthy individuals and merchant banks with money to spare.

13.5 Economic incentives for R & D

One of the most striking facts about science is that expenditure on R & D varies greatly from industry to industry. Industries founded on *science-based technologies* (§9.2), as in the manufacture of aircraft, electronic instruments and pharmaceuticals, usually have a strong tradition of heavy spending on R & D. Firms may be re-investing as much as 20% of their turnover in the invention and development of new products and processes, giving rise to very rapid technical innovation and change. The exponential increase in the performance/cost ratio of solid-state electronic components since the invention of the transistor in 1948 indicates the speed with which technological change can take place when R & D is a major factor in commercial competition. But there are other major industries, such as building and food processing, which have not been in the custom of investing their profits in R & D, and which may spend no more than 2% of their turnover in this way. In such industries innovation proceeds at a much more leisurely pace, despite the possibilities for technical improvement that seem to be on offer to the energetic entrepreneur.

To the external observer, there seems no *a priori* reason why there should be such differences in R & D expenditure from industry to industry. The explanations that would be offered in each case would refer to the technological history of the industry – house building, for example, goes back in an unbroken tradition into the mists of antiquity – or to the distribution of large and small firms in the business, or to overwhelming external influences such as the demand for superior performance in warlike weapons. In general, however, R & D expenditure in a particular industry is maintained at a characteristic level by convention and imitation within that industry rather than by any convincing economic rationale. Why, for example, does the aircraft industry spend much more heavily on R & D than the automobile and railway industries? To say that aircraft manufacture is more 'progressive' and 'technically competitive' simply begs the question why these general characteristics should have become established in these very comparable human activities.

The profit incentive of the free market economy does not, therefore, generate a well-balanced mixture of R & D activities. This is evident in various ways:

> Calculations of the profit likely to be derived from an investment in R & D are so uncertain that most firms simply follow the pattern customary in their industry, regardless of whether this is optimal for the product or service they are providing.
>
> There is no financial incentive to undertake basic or strategic research whose outcome is unlikely to be of use except as a public framework of reliable scientific knowledge. A free market system has no mechanism for the support of science in the academic mode, which should constitute at least a few per cent of total expenditure on R & D.

Since any calculation of costs, benefits and profits into the more distant future – e.g. beyond about 20 years – is completely uncertain or may even be discounted almost in zero terms of present value, efforts directed towards gaining such benefits lie beyond the accountancy horizons of individual capitalist firms. Thus, there is very little incentive for R & D targetted towards the satisfaction of long-term needs, such as the exploitation of renewable sources of energy.

Many obvious social needs and benefits (§12.1), such as those connected with agricultural productivity, public health care and environmental quality, are not easily related to corporate profits. Private enterprise cannot be trusted to undertake R & D directed towards meeting such needs or providing such benefits, which must therefore be fostered by communal institutions such as governments and charitable foundations.

Because the profit motive cannot be relied on to produce all the science we need, only about half the R & D expenditure of most advanced capitalist countries is undertaken by industrial corporations in the private sector of the economy. The major part of this expenditure is for technological development, merging into engineering design and product testing in a few science-based industries, often closely involved in the production of military hardware. The sums that eventually get spent on more basic, longer-term or more socially relevant research are often very large, but they have to come from the public sector of the economy (§14.1) and can seldom be formally justified by a quantitative analysis of relative costs and benefits.

Further reading for chapter 13

Elementary accounts of most of the topics covered in this chapter are to be found in

K. Norris & J. Vaizey, *The Economics of Research and Technology*. London: George Allen & Unwin, 1937

F. R. Bradbury, 'The Economics of Technological Development', in *A History of Technology*, ed. T. I. Williams, Vol. VI. Part I, pp. 48–76. Oxford: Clarendon Press, 1978

C. Freeman, 'Economics of Research and Development', in *Science, Technology and Society*, ed. I. Spiegel-Rösing & D. de Solla Price, pp. 223–76. London: Sage, 1977

The difficulties of assessing the value of R & D are shown by

K. J. Arrow, 'Economic Welfare and the Allocation of Resources for Invention' (pp. 141–59); and K. Crossfield & J. B. Heath, 'The Benefit and Cost of Government Support for Research and Development' (pp. 208–23); in *Economics of Information and Knowledge*, ed. D. M. Lamberton. Harmondsworth: Penguin, 1971

Instructive articles on the R & D leading to technological innovation are

J. Schmookler, 'Economic Sources of Inventive Activity'; (pp. 117–36);

R. Nelson, 'The Simple Economics of Basic Scientific Research' (pp. 148–63);
Z. Griliches, 'Research Costs and Social Returns: Hybrid Corn and the Economics of Innovation' (pp. 211–28); in *The economics of technological change*, ed. N. Rosenberg. Harmondsworth: Penguin, 1971

A survey of case studies on invention and innovation is given by

D. Jewkes, D. Sawers & R. Stillerman, *The Sources of Invention*. London: Macmillan, 1961 (pp. 169–93)

and

J. Langrish, M. Gibbons, W. G. Evans & F. R. Jevons, *Wealth from Knowledge*. London: Macmillan, 1972 (pp. 1–190)

A summary, with references, of the 'Hindsight' and 'TRACES' Projects is given by

E. Layton, 'Conditions of Technological Development', in *Science, Technology and Society*, ed. I. Spiegel-Rösing & D. de Solla Price, pp. 197–222. London: Sage, 1977

An analysis of the theory of 'long waves' is given by

C. Freeman, J. Clark and L. Soete, *Unemployment and Technical Innovation*. London: Pinter, 1982

14

Science and the State

'I wished to procure for science some right to take the initiative in public affairs.'
Werner Heisenberg

14.1 Government support for science

There is nothing new about State support for science. From the seventeenth century onwards, scientists have been directly employed as government officials to chart the land, the seas and the skies, to check weights, measures and coins, to supervise the manufacture of dangerous chemicals and explosives and many other technical jobs. The industrialization of society as a whole has merely enlarged the responsibilities of every government for the welfare and security of its citizens, and correspondingly increased the scale and sophistication of the scientific work that has to be done by the government apparatus (§10.6).

Government *patronage* of 'pure' science also goes back a long way into history. In Britain, the Royal Society and other learned societies were institutionally independent of the State, but were sufficiently close to the centres of authority to extract occasional subsidies for major scientific projects (§10.5). The absolute monarchies of France, Prussia and Russia went much further, by setting up national academies whose members were paid a personal stipend to do full-time research (§10.3). Whatever the level of financial patronage it received, pure science was valued by the State as a cultural ornament, a sign of national superiority, and as a potential source of economic and military benefit.

Nevertheless, although the academic scientific community was never averse to receiving government support for its larger projects, and many scientists were glad to have government employment, there was always a feeling that it should not become too dependent on the State for fear of losing its intellectual autonomy. In recent years, this situation has decisively changed. The scientific activities of most countries are now very largely financed by their central governments, and most scientists are, in effect, employees of the State. In socialist countries, of course, all

R & D organizations, from the most academic to the most technological, are organs of the State apparatus. But even in capitalist countries such as the United States, where private corporations spend a great deal on industrial research, about half the total expenditure on R & D now flows through the federal budget. Government support for science is particularly important in the less-developed countries, whose local private industries seldom have the resources to undertake R & D on their own account (§13.4) and usually have to rely on foreign multinational corporations for scientific and technological know-how.

This development was inevitable, for elementary economic reasons (§13.5). Only the State can find the resources for 'Big Science' projects, running to tens or hundreds of millions of pounds, with no realistic prospect of any financial return (§11.4). Only the State can feel sufficiently confident of its permanent existence to take on very-long-term research projects relevant to, say, the maintenance of energy supplies or the preservation of natural species (§13.5). Above all, the State has communal responsibilities such as national defence, public health and social welfare, which can only be met, in the long run, by a proper mix of basic, strategic and targetted research (§12.2). Although the relative balance between private and public financing of R & D may be shifted one way or another according to the political and economic theories of the party in power, there seems no way back to a world where science is not vitally dependent upon public funds.

The financial dependence of modern science on the State is, of course, one of the salient characteristics of the overall *collectivization* of science in the past half century (§11.5). It has inevitably generated a variety of administrative connections between R & D organizations and the machinery of government, linking back and forth across one of the major interfaces between science and society. These relationships of authority and accountability transcend the conventional boundaries between the academic disciplines of sociology and political science: it has become impossible to give a satisfactory account of the external sociology of science without reference to some of the theories and practices of national politics which make themselves felt deep within the world of research.

14.2 The politics of science

Contemporary society is so permeated with science that many general political issues are really connected with scientific questions (§16.5). But the main topics that get discussed under the heading of *science policy* are quite specialized, and do not loom large on the political scene. The headlines under which political journalists write on this subject fall into three main groups.

Resource allocations; i.e. who gets how much for which sort of research. Typical headlines might be 'Britain witholds contribution to budget for international

nuclear centre' or 'Congress gives go-ahead for R & D on new missile system'.
Administrative arrangements; i.e. which department or agency manages which
R & D activities: e.g. 'Engineering to have separate Research Council' or 'National
Academy of Sciences advises reform of Federal Government Laboratories'.
Technological projects; i.e. what costs and benefits will arise from which plans and
proposals: e.g. 'Britain to build five nuclear power stations of advanced design'
or 'Japan developing new generation of computers'.

But the last of these themes, although of the greatest importance in the real world, is only incidentally concerned with science as such. More and more issues concerning the use and abuse of advanced scientific technologies are coming to prominence in the political arena of every country in the world, but these involve so many other economic, social and ethical considerations that they cannot be dealt with satisfactorily in their metascientific aspects alone.

Since this book does not pretend to be a general text on 'science, technology and society', we restrict ourselves here to a study of the direct financial and administrative relationships between science and government. Even within this restricted definition, science policy is a complicated subject. To understand it in practice, one must have a good knowledge of the political and governmental system of the country where it is made and carried out. Science policy is *policy*, not science, and obstinately refuses to conform to universal principles. Nevertheless, certain types of problem are met with in all countries, and must somehow be dealt with by whatever sociopolitical devices are available. Generally speaking, science policy involves the problems of *choice*, *patronage* and *control*. Although these problems are all closely connected it is convenient to consider them separately, in somewhat schematic form.

14.3 Criteria for choice

Decisions on the allocation of resources between competing R & D projects are the basic building blocks of science policy. This problem of *choice* arises at every level in industrial, governmental and academic science; indeed, it is implicit in the notion of decision-making in all human affairs. Shall we construct a proton accelerator or an electron accelerator to look for zeta particles? In the 'war against cancer' should we give priority to research on viruses or on environmental carcinogens? Should we buy more tanks for the army, or more ships for the navy? Whatever considerations may govern policy in principle, this is the form in which policies have to be put into practice.

Such decisions are particularly difficult in relation to R & D, because most R & D projects, however cut and dried they look on paper, are essentially uncertain. A research project, by definition, is only worth undertaking if it has a real chance of failing (§3.7): it should always look more like a gamble than a safe investment (§13.4).

This risk factor, moreover, is open-ended. The significance of success will not be clear in advance, so that any of a number of other possible investigations can seem equally attractive. The task is not made easier by one of the characteristics of a good scientist (i.e. one whose proposals are worth supporting) – the imagination and enthusiasm to think up many more excellent research projects than can possibly be undertaken.

One of the major questions in the theory and practice of science policy is whether there are any general *criteria* by which such decisions could, or should be taken. As we have seen, economic criteria are appropriate in the final stages of technological development (§§13.3–13.4), although the canons of financial accountancy are seldom strictly applicable. But quantitative calculations of costs and benefits are totally unreliable in relation to basic and strategic research, and provide little guidance in mission-oriented R & D relevant to non-economic benefits such as health and national security. Are there any unquantifiable but rationally ordered principles by which to choose between comparable scientific projects?

The *criteria for scientific choice* proposed by Alvin Weinberg in 1963 fall into two sets. For *external* criteria one should try to answer the following questions:

Does the research have potential applications of great *social* value?

Does the research lead to fairly obvious improvements of existing or proposed *techniques*?

Would a development in that field have important consequences in *other* fields of science?

The *internal* criteria are suggested by questions such as:

Is the particular topic ripe for exploitation?
Are there good things to be done in that field?
Is the subject not too stale and overworked?
Are good people available to do the research?

The Weinberg criteria are obviously rather vague, and have not proved to be a precise instrument by which to guide science policy in practice. But they are a valuable checklist of the considerations that actually enter into R & D decisions, and encapsulate the issues that arise when hard choices have to be made. In particular, it is interesting to see how the relative weight to be attached to these various questions changes as one moves along the spectrum of relevance (§12.2).

The first two external criteria – *social merit* and *technological merit* – are evidently dominant in mission-oriented research and technological development. These criteria, which would be quite intelligible in principle to any non-scientist, effectively define the 'applied' end of the spectrum. To make wise choices by these criteria is the essential art of research management (§12.4), which is thus comparable to the risk-taking skill of the commercial entrepreneur or politician, who must try to reach certain specified objectives with limited information under changing conditions.

But the essence of strategic and basic research is that these two criteria of external relevance are *not* operative. Strategic research may be fostered over a whole field of knowledge, such as plant genetics, with some confidence that particular social or technological benefits will arise from it, but that is not a useful criterion for choice *within* that field. The prime characteristic of a *basic* research project is that it is not directed towards such benefits at all, and thus has no definable social or technological merits in itself.

The notion of *scientific merit* is somewhat different from the other two external criteria in that it puts value on the progress of the scientific enterprise as a whole. Knowledge production is thus held to be a meritorious activity in its own right (§16.6), without reference to the sociotechnical context in which it takes place. At first sight, this seems simply to proclaim the academic ethos of doing pure science 'for its own sake' (§§6.3, 10.7). But it is really a valuable corrective to the traditional individualism of academic scientists, who become so specialized (§5.3) within their own little problem areas that they tend to regard other scientific fields as 'external' and irrelevant to their research plans. In reality, established scientific knowledge is a loosely woven network of facts and theories (§3.8), whose validity and further development is vitally dependent on the cross connections that can be made between specialized fields of research. General scientific merit is thus a very important criterion of choice in policies for basic research, where it is all too easy for each scientist to restrict the search for knowledge to investigations lying within his or her recognized field of competence, regardless of what other scientists are doing in neighbouring fields.

The internal criteria, on the other hand, are precisely those that would be taken into consideration by an individual academic scientist, weighing up the pros and cons of a personal research project. These considerations might not be made explicit, but would involve such factors as the amount of time and other resources that would need to be assigned to the research, the likelihood of discovering something important, and the recognition that might be the reward for such a success (§5.1). These factors, in turn, could only be weighed up in the light of understanding of the validity of the current paradigms in the field (§7.5), the 'state of the art' in research techniques, and the research programmes being undertaken by fellow members of the invisible college of the subject (§5.4).

The problem of choice in basic science, and in most areas of strategic research, is not simply that expertise is needed to assess the likely outcome of a research project: it is that the *significance* of that outcome may only be apparent to experts within that field. In their famous paper on the structure of DNA, Crick and Watson hinted that they fully appreciated the significance of their biochemical discovery in other sciences such as genetics: there can only have been a hundred or so people to whom that would have been obvious at the time, and who could therefore have made a case for giving priority to further work on this project, above other work in the same general field. One of the fundamental difficulties of science policy, spreading into the problems of patronage and control, is that the primary criteria for choice in any particular field of basic and strategic research can scarcely be made intelligible to anyone except a research worker in that field with long experience of making such decisions and living by their consequences.

14.4 The dilemma of patronage

When politicians and civil servants add up the sums spent by the State on science and technology, their attention fastens on the large fraction that goes into mission-oriented research and technological development. The modern State incorporates a variety of R & D organizations to conceive and develop warlike weapons, to protect public health and the natural environment, to support industry and commerce, to foster agriculture, forestry and fisheries, to improve housing and transport, and to monitor education, justice and other social services. To a large extent the problems to be studied by such organizations are defined by the circumstances of real life, and R & D designed to deal with them can be planned and managed in the traditional style of applied science (§12.4).

It is generally agreed, however, that strategic and basic research must also be supported in due proportion. Just what that proportion should be is a question that eludes a convincing analytical answer, although it is often said that something like 10% of the total R & D budget should be allowed for research that is not directed towards the solution of specific problems. Given all the uncertainties of scientific progress, this may be sound enough as a rule of thumb; but it does not help the practical politician who has to decide where this money should come from and to whom it should be handed out.

State support for basic science is caught in the dilemma of *patronage*. The traditional definition of academic science (§10.7) is that it is done 'for its own sake'. In practical terms, this can only mean 'whatever research the scientist(s) think(s) worth doing'. It implies autonomy in the choice of research problems for individual scientists or the heads of research teams. If this principle is accepted, the State can only take part as a benevolent 'patron' of science, providing funds to a special group of people without any strings except some vague promise to 'further knowledge of basic principles' (§11.1). But the State cannot behave like an idiosyncratic millionaire; patronage of science and other individualistic cultural activities is incompatible with the accountability for actions and expenditure that is a hallmark

14.4 The dilemma of patronage

of responsible government. And yet, another hallmark of responsible government is forethought for the collective welfare over the immediate horizons of action, which means strong support for basic science (§13.5).

This problem was not very serious when State support for pure science was episodic and perfunctory. The leading members of the scientific Establishment (§§5.6, 10.5) could lobby their cronies in the political Establishment, and sometimes collect a subsidy for some favoured project. But now that almost all basic and strategic research is financed directly or indirectly from State funds, these matters cannot be settled by informal backstairs methods. The question whether Professor X's supersynchrotron will produce a better scientific dividend than Professor Y's linear accelerator involves hundreds of millions of pounds, and quickly becomes an issue of public policy. As we have seen, the only competent judges of this are Professors X, Y and Z (who has his own project to advocate) all of whom are in competition for the funds at issue. Any decision on this matter by non-specialists would be technically inept and inefficient; yet it seems an abnegation of policy to hand over such large sums of money to such a small closed group without exercising some influence on how they are to be used.

The central problem of State patronage for science is thus to set up an administrative structure in which the internal criteria for choice (§14.3) can be given adequate weight without losing sight of the external criteria, and of other social, economic and political considerations. The conventional practice is to devolve such decisions down through successive layers of quasi-independent committees, such as research councils, 'boards' and 'panels', to the level of a group of people with enough specialized knowledge to understand the technical arguments in an application for a research grant, and in the reports of expert referees (§4.5) asked to assess its merits. Although the group normally consists of scientific authorities in relevant subjects, they are not given executive power to distribute the funds just as they like, but may only respond to the proposals that come to them, and must justify their decisions formally to the committees above them in the funding organization.

This *peer review* system is thus very different from a managerial bureaucracy. It ensures some sort of practical compromise between the personal autonomy of the scientist applying for a grant and the public accountability of the grant-awarding organization. By allowing room for individual initiative in formulating research proposals and in the day-to-day performance of research, it preserves the form, and much of the substance, of the academic ideology (§6.3). But it puts into the hands of a group the vital power to decide whether or not a particular investigation should actually be undertaken, and is thus a major device for the collectivization of basic science (§11.5).

The distribution of funds for basic research by peer review of competing proposals is firmly institutionalized in many countries. But although it resembles the well-established technique by which scientific papers are chosen for publication (§4.5) it is really a much more delicate and responsible task, with far greater influence on the direction and rate of scientific progress. The fact is, quite simply, that it is much easier to abort an unappealing research project before it gets started than it is to strangle it when it has begun to produce undesired results. Moreover, the outlets for publication in an open society are manifold, and are not easily subject to total censorship by an established scientific élite. But the parallel argument for a plurality of funding agencies, able to take independent decisions on whether to support a specific research project, is much more difficult to put into practice. It quickly runs up against the natural tendency towards administrative tidiness in the machinery of the State and the presumed diseconomy of having the same sort of research being done in two different institutions supported by two different agencies.

The collectivization of science has sharpened the dilemmas of State patronage of basic science. These dilemmas are not completely resolved by the peer review procedure, whose strengths and weaknesses have become the centre of much attention in the politics of science. Many diverse metascientific themes come together at this point. It is a matter for serious philosophical consideration whether genuine exploratory investigations, with all their possibilities for serendipitous discovery (§2.5), can be initiated in this way. It is also questionable whether research results can be satisfactorily validated when critical competition may be inhibited to avoid 'wasteful duplication of research' (§3.7). Peer review panels behave sociologically as central committees of their invisible colleges (§5.5), where current paradigms may be overemphasized (§7.5) and the Matthew effect (§5.5) may be significantly enhanced. There are obvious practical difficulties in observing the norm of disinterestedness in such circumstances (§15.2), and much heartsearching may take place over the cognitive norms of originality and scepticism (§6.2, §12.5). The impossibility of formulating a rationale for such decisions (§14.3) is emphasized by the sociologists of scientific knowledge on the one hand (§8.3) and by traditional academic individualists on the other (§6.4), whilst enthusiasts for citation analysis (§4.2) search for quantitative indicators that might provide some guidance in practice. The procedure as a whole has now become a vital element in the management of R & D organizations (§12.4), in the economics of invention (§13.3) and in many other aspects of science policy (§14.2).

14.5 The limits of control

A major thesis of Marxist metascience is that science and technology are 'superstructures' whose form and content are governed by the class relations in society. According to this thesis, knowledge advances in a manner that serves the material and ideological interests of the ruling class. Historical evidence can certainly be produced in support of this thesis, although more as a general tendency than as a decisive force (§8.1, §9.6). Indeed some Marxists have taken the contrary view that science is so closely linked with practice (§3.2) that it should be considered part of the material basis that actually determines the class structure at each historical epoch.

This undecidable doctrinal issue suggests a much more practical question: to what extent can the direction of progress in science and technology be consciously influenced by societal forces. This is clearly the goal of science policy in its instrumental mode (§9.1, §12.1). State support for science is not provided simply to speed up the advancement of knowledge in general. Strategic and mission-oriented research in particular fields is given high priority in the confident expectation that knowledge will be generated that will help solve particular problems or meet particular needs. Thus, for example, a vast amount of R & D is deliberately encouraged on topics that are relevant to the production of warlike weapons of mass destruction because these are regarded as social necessities by the military authorities of most advanced industrial nations – both capitalist and socialist. Similarly, biomedical research is pushed ahead in directions that favour the interests of well-to-do people in affluent countries, with much less attention to the equally interesting diseases that afflict the poor in the less-developed countries of the world.

There is no doubt of the successes that have been achieved by the deliberate application of scientific knowledge and methods in various spheres of life. Many people go further, and believe that any practical problem can be effectively solved if one puts enough R & D effort into it (§16.3). Following this line of thought, political leaders and commentators assign a very narrow rôle to science policy in the general political scheme. Its task is simply to draw up a list of the most salient problems of the nation, arranged in order of priority, together with an organization chart for an R & D system designed to solve these problems in the corresponding order. In a socialist country with a centralized national plan, the R & D system will be similarly centralized. In a mixed economy, where many human needs are presumed to have been detected and satisfied by commercial firms with their own R & D facilities, this organization chart will not be so orderly, since it will have to cover the R & D activities of the various government departments and agencies whose responsibility it is to tackle various societal problems in an uncoordinated fashion. Whatever the system of government, however, the focus of debate on science policy is on the *ends* to be achieved by research. Argument then centres on the relative values and urgency of these ends in the general welfare or in the fate of the nation. We are urged to spend more on medical than on military R & D, because health is more to be valued than war, or more on strategic research relevant to biotechnology than on research relevant to steel-making because biotechnology is the industry of the future – and so on.

But this conception of science policy completely subordinates it to the general political issues of its place and time. It also assumes that R & D can actually deliver the promised goods. In effect, it is supposed that the future pattern of scientific discovery and invention can be consciously controlled by the allocation of resources to chosen sectors of the R & D system.

This supposition is not, by any means, fallacious. A determined R & D effort can usually make some progress in clarifying the nature of a problem, and suggesting some of the paths along which a solution might be found. But this process cannot be forced. Where the 'problem' involves human behaviour, there may not even be a conceptual framework within which it can be located and defined: social scientists are the first to admit the limitations of their knowledge when it comes to social action (§16.5). Even in fields which seem straightforwardly material and practical, there are distinct limits to the deliberate control of science and technology by public policy.

These limits are most evident in basic science, which evolves by the initiatives of individual scientists, mostly working in narrowly specialized problem areas, concerned with what results might be obtained to solve their cognitive problems, rather than with what knowledge is desirable for other reasons (§7.1). The knowledge that is already available is so vast, so abstruse in detail, and yet so interconnected epistemologically (§3.8) that it is simply impossible to calculate where to put the extra effort in order to achieve a desired outcome. As the failure of President Nixon's 'War on Cancer' demonstrated, this process cannot be hastened unless the basic knowledge to be put into practice already exists in exploitable form.

Strategic research looks more amenable to planning and control, but its actual results are often very different from what was originally imagined. A great deal of mission-oriented R & D also proves irrelevant to its supposed purpose because of the difficulty of defining problems and designing specific research projects to deal with them. This is shown in practice by the ineffectiveness of the 'Rothschild' procedure by which various British government departments were to be the 'customers' who commissioned research from 'contractors' - e.g. various research council establishments - in order to solve particular problems of health, agriculture, environmental protection, and so on.

The best that science policy can surely achieve is to accelerate or brake the final stages of development that lead up to the introduction of a technological innovation. This is not an insignificant achievement. Thus, it was massive government support for the relevant R & D that brought a nuclear power industry into being in several countries, before there was any economic incentive for it. On the other hand, renewable energy supply systems, such as solar electricity generators, languished for several decades because they did not seem worth developing during the era of cheap oil. Serious attention to the potentialities of immature technologies is now a major duty of all responsible governments.

Nevertheless, this is very far from control over the technological future. As the story of energy R & D shows, science policy is a clumsy instrument which takes a long time to produce results. Not only will those results be very different from what one had intended, but economic and social circumstances may have changed completely by the time the policy comes to fruition. Some people believe that R & D policies can be guided more effectively by institutionalized *technology assessment*, but there is no convincing evidence that more sophisticated analysis can overcome the long-term unpredictability of technical change (§9.6).

In other words, even the most powerful and well-organized State can exercise only limited control over the direction of scientific and technological progress. 'S & T' must be regarded as an autonomous factor in society, not specifically subordinate to any social class or material interest, although capable of being exploited by any power group for limited periods. This perception of R & D as a potential source of power is implicit in many discussions of the place of science in the political sphere. In a healthy democracy, for example, the policies of State R & D organizations cannot be insulated from public opinion, parliamentary questions, congressional committees, and other lay influences, however inexpert and incompetent these may be in technical terms.

The conventional conception of R & D as an instrument in the hands of some external authority is thus dangerously misleading. In certain circumstances, policy does not control science; on the contrary, science may control policy. This is evident, for example, in military R & D, where the development of a weapons system is highly interactive and reflexive between the scientific and political domains. Sometimes the technical capabilities of a novel weapon cannot be tailored to perceived military demands or supposed military needs: it may be so novel, or have such large effects beyond the battlefield, that the whole military and political framework has to be changed to accommodate it. Thus, the neutron bomb may have seemed, as first sight, like the scientific solution to the tactical problem of defence against a mass of tanks: as it was developed, it became a political problem by bringing into question the underlying assumptions of strategy, diplomacy, and national security.

14.6 Science in government

General political theory casts science as an instrument of political power, and thus gives it a subordinate part in the action. But scientific knowledge, scientific methods, scientific personnel and scientific organizations have come to have great influence in governmental decisions and other public affairs. To what extent should 'Science' be considered a distinct institutional factor in national life, on a par with such traditional institutions as the Army, the Law and the Church.

The scientific community has never been alienated from the State, and has seldom shown more than token opposition to the policies of the government of the day.

Despite the famous example of Galileo's defiance of the Church, scientists have almost always proved themselves loyal – indeed devoted – servants of whatever body controlled the machinery of power. This was clearly demonstrated in the Second World War, when the scientists of each nation gave their services without stint to the national cause, whether in the name of democracy, communism or nazism. Since the time of Sir Isaac Newton, who combined the official post of Master of the Mint with the Presidency of the Royal Society, the leading dignitaries of the scientific community in every country have been publicly honoured and rewarded by their governments, and occasionally played an important part in government affairs (§10.5, §16.1). In some countries, scientific eminence has been publicly 'recognized' by the award of high official rank in the State bureaucracy, with all that this might imply of authority inside and outside science (§§5.7).

The traditional élite of the scientific community have not, therefore, been openly opposed to the collectivization of science (§11.5), and the virtual *incorporation* of large parts of it into the State apparatus. In this process, of course, many of them have been given considerable managerial or administrative responsibilities, as directors of government laboratories, advisers to government departments, chairmen of research councils, and other leading posts in the R & D system. For a few this path has opened the way to some of the highest positions of power in the nation, at the head of a major nationalized industry or large industrial firm. Although it is very rare indeed for an experienced research scientist to be elected to Parliament, or any corresponding legislative assembly in any other democratic country, a number of eminent scientists and technologists in Britain are life peers, and exert an active influence on legislation and government policy. There is no doubt at all that science has moved closer to the power centres of society, and that scientists now form a recognizable segment of the political 'Establishment' in most advanced countries (cf. §§5.6).

Nevertheless, it is only in communist countries that science has been completely incorporated into the State. In countries such as Britain and the United States, the leading organs of the scientific community, such as the Royal Society, the National Academy of Sciences, and innumerable specialist learned societies, are autonomous institutions with a powerful identity. Their collectivization is informal and indirect, through financial subsidies, participation in policy by consultation, and overlapping membership of bureaucratic hierarchies and committees. At the moment it is hard to say whether these traditional – and highly charismatic – institutions are losing power to the official machinery of government R & D, or whether, on the contrary, they are the medium through which this machinery is informally directed.

This issue could obviously be resolved in various ways, depending on the national political circumstances. But the question remains whether science is on the way to becoming one of the 'Estates' of the polity, with a distinctive communal influence

14.6 Science in government

in public affairs. Until recently, it was too weak and peripheral to have a noticeable presence in the general political arena. Now, with something like 1% of the population and material resources of the nation, it could have a significant influence, if it were to speak with a single voice. Many campaigners for radical political causes, such as unilateral nuclear disarmament or world government, have noted this, and tried to get 'Science' on their side.

This image of science as a unified estate, comparable, say, with the Church of England or the Navy, is far from reality. The notion of a 'scientific community', governed by a coherent élite, was always exaggerated and is certainly not true today (§§5.6, 6.4). The individualism of the academic ideology still speaks forcefully for the autonomy of universities and other academic institutions. The various segments of collectivized science – academic, quasi-academic, governmental and industrial – are not organized into a national 'R & D system' under central command. This is not necessarily a weakness or an historical accident to be remedied by reform. There is a well-founded argument for a Minister of Science, to formulate and carry out an integrated science policy over all aspects of natural life; there are equally persuasive counter-arguments for putting most R & D under the control of the separate departments, agencies and firms that will apply its results (§14.5). Although some Prime Ministers and Presidents have made good use of a chief scientific adviser, who has then functioned as the effective head of the whole R & D community in the country, others have managed well enough without any centralized apparatus of scientific consultation and control.

At this level, the politics of science is just the same as politics in general. If, as in the United States, federal policy is the outcome of a public power tussle between quasi-autonomous agencies, congressional committees and presidential aides, then the scientific estate will be similarly fragmented and factious. If, as in Britain, government policy is made behind the scenes by the 'old boy network' of senior members of the Establishment, then that will also describe the organization of the R & D system and its place in the centre of things. If, as in France, it is customary for government policy to be formulated and carried out by a rigid framework of bureaux, then the affairs of science will be the concern of some sub-set of these bureaux. If, as it is alleged of many countries, there is a 'Military–Industrial Complex' with too much power for the good of the nation, then it is likely that science will have to ally itself with this lobby, and be named as a member of an even more sinister 'Military–Industrial–Scientific' trio. And if, as in the Soviet Union, the monolithic power of the State is a façade for a struggle for power between factions entrenched in various government and party institutions, we may be sure that the Academy of Sciences, a highly esteemed and relatively coherent institution with direct responsibility for much of the R & D effort of the country, is also a seat of general political power for its leading officials, and is thus of some account in national affairs. Every country, as the saying goes, gets the government it deserves – and nowadays science is thrown in with government, for good measure.

Further reading for chapter 14

The machinery of the government of science in Britain is described by
> P. Gummett, *Scientists in Whitehall*. Manchester: Manchester University Press, 1980 (pp. 20–53, 214–237)

Basic documents on British science policy are collected in
> J. B. Poole & K. Andrews (ed.), *The Government of Science in Britain*. London: Weidenfeld & Nicholson, 1972

Brief accounts of the 'science systems' of the United States and the Soviet Union are given by
> M. N. Richter, *The Autonomy of Science*. Cambridge, Mass: Schenkmann, 1980 (pp. 79–130)

The 'criteria for scientific choice' are given in
> A. M. Weinberg, *Reflections on Big Science*. Oxford: Pergamon, 1967 (pp. 65–100)

The problem of control is discussed by
> R. Johnston & T. Jagtenberg, 'Goal Direction of Scientific Research', in *The Dynamics of Science and Technology*, ed. W. Krohn, E. T. Layton & P. Weingart, pp. 29–58. Dordrecht: D. Reidel, 1978

and by
> H. Brooks, 'The Problem of Research Priorities'. *Daedalus*, Spring 1978, (pp. 171–90)

The rôle of science in politics is analysed at length by
> D. K. Price, *The Scientific Estate*. Cambridge, Mass: Harvard University Press, 1965

The political and economic background of military R & D is described by
> H. M. Sapolsky, 'Science, Technology and Military Policy', in *Science, Technology and Society*, ed. I. Spiegel-Rösing & D. de Solla Price, pp. 443–71. London: Sage, 1977

15

The scientist in society

'When you see something that is technically sweet, you go ahead and do it and you argue about what to do about it after you have had your technical success. That is the way it was with the atomic bomb.' *J. Robert Oppenheimer*

15.1 Towards a social psychology of science

Science is what scientists do. The scientific life is notorious for the demands it makes on the mind and on the spirit. The social psychology of science is thus an essential metascientific discipline, along with philosophy, sociology, politics and history.

The traditional academic ethos (§6.3) lays great stress on the individuality of scientists, and thus emphasizes the distinctive mental and emotional traits that tend to separate them from the mass of people and from one another. The naïve history of science is a chronicle of heroic or saintly *personalities* who have triumphed through their innate abilities and virtues. More seriously and soberly, psychologists have tried to delineate or discover the personality types that are characteristic of scientists in general, or of scientists in particular disciplines, such as theoretical physics or experimental biology.

Unfortunately, these investigations have not proved very conclusive. Mature scientists, and even science students, are not, presumably, 'just like everybody else', but careful empirical research on their personality traits has not provided reliable insights that are superior to everyday 'folk' understanding of these matters. Scientists have to be reasonably intelligent in a conventional sense – if only to master the academic knowledge they need to get to the research frontier – and they have to be well motivated to maintain individual initiative in research. But it is not clear how their successes and failures are really associated with deeper forces and factors, such as intellectual convergence, introversion, indifference to authority, sublimation of neurotic tendencies, etc., as advocated by various schools of individual psychology. If anything, these studies simply draw attention to the bewildering variety of personality types to be found amongst scientists, and to the uniqueness of every scientist as a person.

From our present point of view, however, the social psychology of science is complementary to its sociology. Academic science, for example, is not anarchical, in spite of its individualism. The traditional scientific community functioned through the rules and norms that its members had learnt to follow (§6.1). The collectivization of science has changed many of these rules (§12.5), but contemporary scientific institutions are not machines: they operate by the coordinated actions of people who are conscious of what is expected of them as events unfold around them, and who voluntarily perform as expected. The personalities of scientists as individuals are thus less significant than the *rôles* they are called on to play as members of these institutions.

Laboratory life is not, of course, precisely scripted and stereotyped. If one studies it in detail, one finds as much diversity in the parts that people play as in the personalities that colour their actions. Indeed, as we all know, there is no way of separating the social rôle from the person who plays it – just as there is no way of separating nature from nurture in the making of each individual. The significant point is, rather, that any one scientist may have to play several different rôles, depending on the social group or circle in which he or she happens to be situated. Most people have this experience in relation to family life, on the one hand, and working life on the other; the rôle of the parent, for example, is sharply differentiated from the rôle of, say, the customs official, even though the same person performs both these rôles in the course of a single day. A salient feature of contemporary scientific life is that it can no longer be experienced as the performance of a single vocational rôle. Quite apart from his or her normal rôles as a member of a family and as a tax-paying, property-owning, law-abiding citizen, the modern scientist may be called upon to play several distinct professional rôles, within the world of science or in society at large. A brief account of these rôles will give some idea of what is involved when we talk nowadays of 'the place of the scientist in society'.

15.2 The scientist as intellectual entrepreneur

The individualism of academic science casts every scientist in the rôle of an *intellectual entrepreneur*, undertaking research on his or her own initiative on the basis of a personal assessment of the likelihood of making a discovery, and rewarded with personal recognition if successful (§5.1). The traits of character appropriate to this rôle are familiar to every aspiring scientist: *curiosity*, to pick up the first hint of a serendipitous discovery (§2.5); *intelligence*, to grasp contemporary theory and to formulate new research problems (§2.15); *persistence*, to carry through a long and laborious investigation to a convincing conclusion; *honesty*, to validate one's own results objectively (§3.2), and to present them fairly to others (§4.3) – and so on.

These traits have been so idealized and eulogized in the folklore of science that some people have come to believe that they are features of a specific 'scientific attitude' which would work miracles if only it were applied to all the practical problems of life. This naïve view ignores some of the other traits that are inseparable from this rôle, such as *narrowness* of view, to attain mastery of a specialty (§5.3), and *egoism*, to concentrate on a topic, and to prevail against competition (§5.2). These vices of academic life are not apparent in the formal literature of science, which is carefully scripted to conform to the traditional norms (§6.4), but they are probably just as essential to the dynamical stability of academic science as are the virtues that are so widely praised.

To describe the academic scientist as an 'entrepreneur' is to suggest a direct comparison with the rôle of the commercial or industrial entrepreneur in a capitalist society. Scientific work is thus treated as a productive process in which publishable 'contributions' are exchanged in an intellectual market place for 'recognition' (§5.1). This metaphor is sometimes extended further, by applying terms such as 'resources' and 'capital' to the symbolic entities that enter into this process. Whether or not this extension of the metaphor is justified, the comparison is apt: historians of science are generally agreed that the scientific revolution of seventeenth-century Europe was closely linked with the transformation from feudalism to capitalism in the political and economic structures of society as a whole. This transformation was associated with a religious reformation which provided powerful spiritual and moral support for the rôle of the capitalist entrepreneur. The historical connection between protestant theology and the natural philosophy of science is much disputed, but there is no doubt that the individualism of the traditional scientific ethos harmonizes perfectly with the protestant ideal of an 'inner-directed' personality obedient to internalized norms of behaviour and striving towards transcendental goals.

In countries such as the Soviet Union, where science has been fully collectivized (§11.5), the notion of the scientist as an intellectual entrepreneur is repudiated by the official ideology, although it still lingers on as a motivating factor in the lives of many scientists. But in some countries where private enterprise is a dominant economic force, particularly the United States, this feature of the academic ethos is actively reinforced by the funding procedures for basic and strategic research. Even scientists with permanent tenure as university professors have to compete individually for the material resources they need for their work. Like small farmers or shopkeepers seeking business capital from a bank, they must apply to a funding agency for a grant that will pay for apparatus and assistants – even for a proportion of their own personal salaries (§14.4). They are thus forced to play the rôle of the individualist entrepreneur very earnestly indeed, often transgressing the norms of communalism and disinterestedness (§12.5) in their anxiety to survive as active scientists.

15.3 Citizen of the republic of science

The norm of communalism (§6.2) casts the academic scientist as a member of a cooperative community. This rôle thus counteracts the extreme individualism of the legendary 'lonely seeker after truth'. In its idealized 'Mertonian' form, the scientific community functions as a self-governing republic, where every qualified scientist claims the rights and the responsibilities of a free citizen. To perform this rôle satisfactorily, a scientist must therefore be ready to *communicate* research results (§4.3), *cite* the work of other scientists (§§4.2, 5.2), give voluntarily of time and effort as a *referee* or *editor* (§§4.4, 4.5), take part in scientific *meetings* (§4.7), *reward* notable scientific achievements (§5.1), defer to the *authority* of more esteemed colleagues (§5.6), and perhaps, eventually, if fortune smiles on his or her endeavours, bear graciously and wisely the burdens of scientific leadership in a learned society, a university, a government department or an industrial firm.

This rôle is evidently characterized by the social contexts in which it has to be played, rather than by the personality traits of the actor. Many of these contexts, such as those that arise in scientific meetings, are customary rather than formally defined, and call for conventional forms of action that are easily learned, such as praising a very dull speaker for his contribution, or asking a controversial question in an unaggressive tone of voice. Other contexts, such as those associated with the formation and management of a learned society (§7.2) may be more systematically codified, and may demand a good deal of personal initiative and social sensibility. The academic ethos takes for granted that these functions will be carried out by somebody – preferably by a scientist of high research ability – but offers little encouragement to do the job well. The sociological reality is that the progress of science is very dependent on the competence and conscientiousness with which these major rôles are assumed and performed by a relatively small minority of the scientific community (§5.6).

The collectivization of science has affected the communal responsibilities of scientists by bringing the scientific community under firmer external control (§11.5). In countries where the traditional institutions of science have been incorporated in the state apparatus, scientists are now expected to act as state functionaries, rather than as citizens of a 'republic of learning'. In most other countries, academic institutions such as universities and learned societies are expected to account in some detail for the subsidies they receive from the State (§14.1). They are thus obliged to rationalize and codify their practices, and to set up administrative structures where working scientists have much less occasion to act out their traditional communal rôles. Duties that used to be performed voluntarily, such as editing a learned journal or negotiating the stipend of a research assistant, have become so heavy and complex that they have to be put in the hands of paid professionals. This may free the scientists for their real work of research, but diminishes their personal involvement in the affairs of the academic community.

A major characteristic of the scientific community is that it is not, in principle, bounded by national frontiers. In academic science, every invisible college is trans-national in membership (§5.4). Scientists travel to meetings all over the world, and often work abroad for months or years. National academies and learned societies maintain fraternal links, and are formally associated in international unions. In recent years, scientists from many nations have collaborated closely in a number of major research projects, through networks such as the International Geophysical Year, and international institutions such as CERN (the European Council for Nuclear Research) (§11.4). Although the resources for these projects come from national governments or intergovernmental organizations such as the United Nations, the scientists themselves take part as members of the world scientific community rather than as international civil servants or citizens of their respective nations.

But this cosmopolitan tendency is not to be trusted in times of political crisis. There may once have been a time when 'the sciences were not at war', but this has certainly not been true in the twentieth century. In both World Wars, the scientists lined up patriotically to serve their respective countries, and even indulged in propaganda campaigns against their colleagues on the other side.

Calls for international solidarity with all those working for the advancement of knowledge have not been without moral force amongst 'pure' scientists, whether in the name of peace or in the name of human rights. The first Pugwash meetings in the late 1950s played some part in breaking the diplomatic ice of the Cold War, by exploiting the trans-national contacts of scientific notables from East and West. But these are little more than gestures in the face of a world divided into heavily armed camps, where something like a third of all scientific work is connected with preparation for war. The rôles of scientists as members of a transnational community directed towards benevolent transcendental goals is tragically subsidiary to their individual rôles as loyal citizens of particular nation-states (§10.3).

15.4 The scientist as technical worker

Industrial science (§10.6) employs scientists as *technical workers*. Their rôles in governmental or industrial R & D organizations are much the same as those of other skilled employees. They are expected to apply their talents conscientiously in support of the policies of their employers, and to work diligently at the tasks prescribed for them. Their rights and responsibilities are limited by their contract of employment; in general, they are bound to follow the instructions of their organizational superiors and to manage the work of those below them in the spirit of those institutions.

As highly qualified professionals, research scientists are often given much more autonomy in the organization of their activities than other workers. They may be at liberty to come and go as they please, to attend scientific meetings in a personal capacity, and to contribute papers to learned journals in their own names. But despite these superficial resemblances to the academic life style, the rôle in which they are cast is quite different. The personality traits demanded of them are those of the 'organization man' (or woman) whose behaviour has to be closely coordinated with the behaviour of others and calculated to advance the interests of the specific organization to which they belong.

As one moves towards the more 'relevant' end of the spectrum of R & D organizations (§12.2), this rôle becomes more and more the norm. Scientists involved directly in the technological development of new commercial products or military hardware are not essentially different from the engineers, production managers, sales staff or military personnel with whom they work. The notion that scientists have a special social rôle cannot be sustained in these circumstances – which are, in fact, precisely those under which the majority of people with advanced scientific training are now actually employed.

The contradiction between the 'academic' and 'industrial' rôles is the source of many dilemmas in the management of collectivized R & D (§12.4). For example, should young scientists be given 'academic' freedom to choose their own basic research themes, or should they commit themselves to the interests of the organization by being put on to immediate practical problems? To what extent should scientists be permitted to criticize the technical perspectives of their employers, in the name of scientific scepticism? Must some scientists be discouraged from taking an active part in the affairs of the scientific community because of their involvement in secret research? What achievements should be rewarded with promotion – contributions to scientific knowledge or the improvement of productive techniques? Should there be a special career ladder for very able researchers who do not desire, or are not fitted for, high managerial responsibility? These dilemmas arise because the stereotype of the scientist as intellectual entrepreneur is simply not compatible with the stereotype of the scientist as technical worker: these conflicting rôles cannot be harmonized without considerable psychological adjustment and spiritual compromise (cf. §12.5).

15.5 The scientist as expert

In principle, every fact or theory known to science is contained in the public scientific literature (§4.1); in practice, this information is only intelligible to a specialist in the relevant field (§5.3). Whenever a practical question arises where such information is needed, a research scientist will probably have to be called in as an *expert*, not only on the published literature but also on the tacit knowledge that relates to it (§3.3). This expertise may be required in a variety of circumstances. An academic physicist doing research on semiconductors might be hired as a *consultant* by an electronics company. A professor of entomology might be asked to give evidence as an expert *witness* in a court case about insect infestation. A microbiologist might be appointed an *adviser* to a public commission on the regulation of genetic engineering – and so on.

Scientific workers employed by large R & D organizations perform this advisory function as part of their normal duties. Outside consultants hired to give confidential advice are in much the same position as 'in-house' experts in that the opinions they offer are understood to be essentially their own. But a person appearing as a scientific expert in public is usually understood to be speaking on behalf of 'science', and is thus being cast in a much less individual rôle. The scientist acting as a *public* expert is supposed to be simply a medium by which objective scientific knowledge is being brought to bear upon the practical problems of the world.

Where the information required is essentially well-established (§3.8), this rôle presents no difficulties in principle, although it may call for considerable skill in practice. But the questions that arise in technology, law, commerce and politics are seldom posed in the contrived and bounded terms of research problems (§2.15), and almost always call for information that is not validated, or has not even been 'discovered' in a scientific sense. The sheer ignorance of science on many weighty issues is very evident, for example, in the development of nuclear power, where it is largely a matter of conjecture what would happen if there were a major reactor accident.

From a strict philosophical point of view, a scientist faced with such a question should repudiate the rôle of an expert altogether. But that would be an antisocial attitude, since it would effectively deny access to whatever relevant information might have been gleaned in the course of research, however uncertain or controversial it might be. The conscientious expert should then present this information very tentatively, indicating its low epistemological status by, say, estimates of the probability of inferred generalizations (§3.5).

This is a council of perfection, based upon an unrealistic philosophy of science. Scientists cannot cast off their emotional commitments, even in scientific controversies, and are bound to express opinions weighted towards their personal inclinations. It is neither surprising nor shocking that they cannot be relied on individually to play the rôle of perfectly objective advisers. The 'truth' and 'objectivity' of scientific knowledge derive from its collective character (§§8.4, 8.5), and are not inherent in the experiences or notions of any single person.

The difficulty is, however, that the supposedly 'scientific' uncertainties cannot be disentangled from other factors in the situation. The sociology of knowledge (§8.2) teaches that the scientist is playing a part in a social drama, and cannot give advice without reference to his or her personal opinions or interests. Indeed, in law suits and planning inquiries the divergent interests of the technical witnesses appearing for the various parties are so notorious that their credibility as experts is normally put to public test by adversarial procedures such as legal cross-examination. For the same reason, it is desirable to arrange support for research pluralistically (§14.4), so that all the scientists competent to advise on the policies of a particular governmental or commercial organization are not dependent upon it for their employment or research facilities. The academic norm of 'disinterestedness' (§6.2, §12.5) is an ideal towards which most scientists may strive, but the rôle of the scientist as an independent, neutral adviser in public affairs can only be sustained within a social framework in which this behaviour is encouraged and esteemed.

15.6 Social responsibility in science

Science is extraordinarily influential in modern society, and yet scientific work is carried out in laboratories and offices that are far removed from the scene of its applications. The isolation of most scientists from the practical outcome of their research is an inevitable effect of the way in which it functions, epistemologically and sociologically. Until recently, this isolation was actually fostered and celebrated in the ethos of 'pure' science (§10.7), undertaken 'for its own sake', without regard for the consequences. Nowadays, however, it is generally agreed that scientists should endeavour to show some responsibility in their actions, especially where the results are socially destructive through political oppression and war.

How should this ethical notion of 'social responsibility in science' be put into practice? First and foremost, every scientist is an ordinary human being and an ordinary citizen (§15.1); there is no case for denying the normal responsibilities of these rôles just because one happens to be a scientist. Unfortunately, in their rôles as individual intellectual entrepreneurs (§15.2), scientists are bound to ignore the wider effects of their research (which are almost incalculable, anyway) and to follow it wherever it leads. The traditional academic ethos reinforces this attitude, which is entrenched in the charters and policies of many of the institutions of academic science.

Nevertheless, as a member of a scientific community (§15.3), every scientist has some responsibility for the 'external relations' of that community (§10.5), which can no longer be disconnected from other societal structures. National academies, learned societies and universities are institutionally involved in political, commercial and military issues, where the ethical sensibilities of their members can play a very important part. In the role of 'citizen of the republic of science', a scientist can join with others in opposing, say, research on biological weapons, even though this might not be an easy position to maintain as an individual subject to the pressures of personal employment.

Scientists employed as technical workers (§15.4) have usually had to surrender much of their ethical independence to the organizations that employ them. The prime act of irresponsibility in this rôle is to work for an organization – e.g. an industrial firm producing faulty goods – whose activities one deplores. At this point, however, more general ethical and legal issues enter the argument, such as the responsibility of the subordinate carrying out orders from above, or the grounds on which an employee should be permitted or encouraged to 'blow the whistle' on the antisocial acts of his or her employers. Scientists tend to be caught in the dilemmas of this rôle because their job is often to predict or monitor the consequences of corporate policies, and they thus become uneasily aware of the defects of these policies.

It goes without saying that the scientist acting as a public expert (§15.5), or taking a leading rôle in public affairs (§14.6) is in the most sensitive position from the point of view of social responsibility. This applies not only to the actual advice given, or the decisions taken: the scientific notable in this situation often claims to be speaking on behalf of 'science', and thus contributes significantly to the perceived rôle of all scientists in society. Scientific authorities (§5.6) are seldom elected democratically in contested ballots, nor are they usually answerable to the rank and file for the policies they pursue or the opinions they express; a heavy responsibility rests upon them to balance justly the personal, communal and societal considerations that arise in the politics of science and technology.

Further reading for chapter 15

Relevant reviews are
> R. Fisch, 'The Psychology of Science' (pp. 277–318); and S. A. Lakoff, 'Scientists, Technologists and Political Power' (pp. 355–91); in *Science Technology & Society* ed. I. Spiegel-Rösing & D. de Solla Price. London: Sage, 1977

The autonomy of the researcher is discussed by
> W. P. Metzger, 'Academic Freedom & Science Freedom', *Daedalus*, Spring 1978, pp. 93–114

Some grave issues in the history of the 'republic of science' are discussed by
> J. Haberer, *Politics and the Community of Science*. New York: Van Nostrand, 1969

An idea of the situation facing the expert witness is conveyed by
> J. S. Oteri, M. G. Weinberg & M. S. Pinales, 'Cross-examination of chemists in drugs cases', in *Science in Context*, ed. B. Barnes & D. Edge, pp. 250–9. Milton Keynes: Open University Press, 1982

Cases of scientists as 'atomic spies' are described by
R. W. Reid, *Tongues of Conscience: War and the Scientists' Dilemma*. London: Constable 1969

The various rôles played by scientists in a major political issue are demonstrated in J. Rotblat (ed.), *Scientists, The Arms Race and Disarmament*. London: Taylor & Francis, 1982

16

Science as a cultural resource

'Physico-mechanical laws are, as it were, the telescopes of our spiritual eye, which can penetrate into the deepest nights of time, past and to come.'
Hermann von Helmholtz

16.1 Beyond the instrumental mode

Scientific research is undertaken nowadays primarily for its eventual material benefits (§9.1). For this reason, our discussion of the external social relations of science has focused almost exclusively on its instrumental connections through technology. But the influence of scientific knowledge and ways of thought is far wider than the contributions of R & D to industry, medicine, agriculture, war and other typical human pursuits (§12.1). In this final chapter, therefore, we consider science as a general *cultural* resource, with significant societal effects beyond those directly due to technical change.

This is a large and diffuse metascientific theme, which can only be treated very schematically. Science is only one amongst the many elements that go into the making of contemporary culture. These other elements – psychic, political, philosophical, humanistic, aesthetic, religious, etc. – have to be appreciated in their own right and not looked at solely through eyes that have already been 'blinded by science'. *Scientism* (§3.9) is not just a philosophical doctrine; it has its sociological, political and ethical manifestations, which are equally misleading and dangerous.

Consider, for example, the topic of the previous chapter – the scientist's rôle in society. Some enthusiasts for science advocate a greater expansion of this rôle; they assert, in effect, that everything would be OK if scientists ruled. Now it is true that success in scientific work calls for impressive qualities, such as intellectual grasp, openmindedness, persistence and honesty, which might be of great value in a responsible political leader. Some scientists have, indeed, played a major part in political affairs (§14.6), whether through the machinery of government, as in the case of Robert Oppenheimer, or simply through the force of their moral example, as in the case of Albert Einstein. But the personal qualities desirable in those who govern the State is one of the great questions of political theory, going back to Plato. The scientistic view ignores other essential qualities for political leadership, such as sociability, persuasiveness in debate, willingness to compromise, appreciation of the needs of ordinary people, or even ruthless ambition, which are not at all characteristic of the 'scientific attitude' (§15.2). It is generally agreed by political theorists that if the *technocratic* tendency of science were allowed to prevail, it would rapidly degenerate into tyranny. In other words, the experience and attitudes gained in and through science are an inadequate guide to the way in which society works as a whole.

16.2 Public understanding of science

How much science do people actually know? To judge by the questions and answers in television quizzes – very little indeed. Even amongst well-educated people, the most elementary scientific facts, such as the chemical symbol for sodium, or the physiological function of the liver, are regarded as highly technical and 'difficult'. Modern culture depends utterly on science-based technologies (§9.2); techniques derived from scientific practice and concepts drawn from scientific theory pervade everyday life (§9.4); yet few people have a general notion of what is now known to science.

This ignorance is deplored by scientists, who press for action to improve public understanding of science. Yet the machinery for this action is fully established. For more than a century, *science education* has been a major function of the school and university systems of all industrialized countries. By the end of their compulsory period of schooling, most young people have had at least a few courses in the basic sciences. Courses at every level, in every scientific discipline, 'pure' or 'applied', are open to suitably qualified students. There are plenty of opportunities to learn science, for those who want to. Science is also widely *popularized*, through books, magazines, newspapers, radio and television. Some of this material is sensational or opinionated, but one can easily find in the 'media' a solid stratum of scientific information presented skilfully by effective communicators. Nevertheless, for the great majority of people, science is a subject that one might have to learn as part of one's job, but is otherwise regarded as difficult, dull, and best soon forgotten.

All specialist groups deplore the lack of public understanding of their specialty and urge that it should be given greater emphasis in mass education and the mass media. But the case of science is instructive because it illustrates the difference between the viewpoints of 'insiders' and 'outsiders'. The outsider's view is overwhelmingly instrumental (§9.1). The whole purpose of science education is taken to be vocational. Science subjects were introduced into the elementary school curriculum, and technical universities were founded in order to train workers, managers and technical experts for industry. The 'attentive public' for popular science is very limited, except where it touches upon material issues of personal health and safety.

From the inside, on the other hand, science is seen primarily as a conceptual scheme by which observable facts are ordered and mapped (§3.8). The emphasis is not so much on utility, as on the possibilities of discovery and of validation. In the opinion of most scientists, what people ought to be made to understand is the 'scientific world picture', in greater or less detail. They tend, therefore, to structure the science curriculum around the central cognitive themes, with very little regard for their applications in everyday life.

There is thus a serious mismatch between the interests of those who are already inside science, and the motives of those whom they would like to draw in. Most people find great difficulty in getting to understand the conceptual schemes of the sciences, which seem so very unlike the familiar structures of the life-world (§3.9). A few young people are attracted to the idea of discovering new representations of reality: the great majority see this as a relatively fruitless task, irrelevant to their personal lives, and calling for more time and effort than they can spare, whether in formal education or in informal learning. Novel educational curricula on the theme of 'science, technology and society' can encourage students and teachers to bridge this gap, but science remains a distinct sub-culture whose actual contents are practically unknown to all but a tiny fraction of the population.

16.3 Folk science, pseudo-science and parascience

Just because people are ignorant about science does not mean that they lack reliable knowledge on which to base their actions. In every human culture, people know very well when to plough their fields, how to treat minor ailments, or what to make of the behaviour of other people. The ordinary problems of life are dealt with by reference to rituals, rules and maxims which may not have been codified and tested scientifically but which are often sharply observed and based soundly on experience (§9.3). Whether or not we want to call such traditional knowledge 'scientific', it was the original starting line for the development of all our natural sciences and technologies.

In the societies studied by anthropologists, everyday knowledge of the life-world (§3.2) is either taken for granted, or is referred back to a loosely articulated system of legends, myths and religious doctrines. In modern society, however, religion has lost much of its authority in relation to practical knowledge, and the efficacy of magic is doubted. People are ready to follow custom, or a convenient rule of thumb in the minor decisions of life, but in really serious matters they feel they must put their trust in science. When they are gravely ill, for example, they demand medical treatment by the most up-to-date scientific methods.

This faith in the practical efficacy of science is not altogether misplaced, but it can become excessive. What is one to do if the guidance offered by orthodox science is inadequate, or unpalatable – for example, when a disease is said to be incurable? People may then be tempted to turn to other sources, supposedly equally 'scientific', in the hope of more 'helpful' advice.

Even in the most advanced societies, there exists a considerable body of *folk science*, of varying degrees of sophistication, outside the domain of orthodox science. From our present point of view, the most interesting characteristic of this sort of knowledge is that it often claims to be 'scientific', despite the fact that it has not been accredited as such by the scientific community. Such claims are addressed, of course, to the general belief in the superiority of science to religion, magic and other systems of knowledge. Thus, cancer sufferers are induced to dose themselves with 'laetrile', a substance 'discovered' by a man with a PhD who offers an elaborate biochemical explanation for its supposed action. On a larger canvas, T. D. Lysenko presented 'scientific' arguments to justify his agricultural methods, and thus became (with the support of Stalin) a folk hero amongst Soviet farmers, even though his claims were not confirmed by properly conducted experiments.

For this reason, the established institutions of science are very hostile to all forms of *pseudoscience*. This hostility is often well founded. People who esteem science highly but who are ignorant of its contents are easy victims of self-deception, if not outright fraud. Scientists would be socially irresponsible (§15.6) if they failed to take a public stand against practices that they regard as grossly misguided or deceitful.

Nevertheless, scientific opposition to pseudoscience is sometimes carried to unwarranted extremes. For example, the attempts by some astronomers to prevent the publication of Immanuel Velikovsky's harmless nonsense about the history of the Earth aroused the suspicion that science was trying to set itself up as the sole authority on all such matters. This intolerance of deviant opinion infringes the scientific norms of universalism and scepticism (§6.2), and cannot be justified epistemologically. Philosophers simply do not all agree that there is a formal criterion (§ 3.7) by which 'pseudo-science' can be unfailingly distinguished from the genuine article.

At its core, established science is a coherent system of fact and theory, without serious rival as a source of reliable information. Scientists mostly work within an accepted framework of regulative principles (§3.10), and strive earnestly to build up an empirically tested, non-contradictory body of knowledge (§3.8). But the scientific enterprise cannot be closed off from other cultural activities and influences. At any given moment it must take notice of propositions of widely varying credibility, ranging from well-validated observations and theories to the wildest shores of fantasy. Securely held paradigms, like the permanence of the continents, fall into disgrace (§7.4), whilst absurd notions from folk science, like the fall of meteorites from the sky, are found to be justified. Sophisticated practical techniques, such as clinical medicine, rely as much on unexplained maxims and untested rules of thumb as on validated facts and theories.

This is not to say, as doctrinaire sociological relativists seem to suggest (§8.2), that one bit of claimed knowledge is as good as another. Folk science is very seldom as reliable as orthodox science where they both apply. Scientists have every right to express their opinion that some knowledge claims, such as those made for extrasensory perception, are so contrary to established understanding, and are supported by so little evidence, that they should be dismissed as *parascience*. But even at that distance from the centre, the margin of credibility is not altogether distinct. It is a fundamental metascientific principle that there can be no sharp line of demarcation between 'scientific' and 'non-scientific' beliefs in everyday life.

16.4 Academic scientism

The word 'science' has been used in this book in the present-day fashion, as if it referred almost exclusively to subjects such as physics, chemistry, biology and geology, and their associated technologies such as engineering, medicine and agriculture. But this word was originally used to denote any orderly body of knowledge or recognized branch of learning. In most European languages the corresponding word – *la science, wissenschaft, scienza, nauka*, etc. – still has this general meaning. Thus, to say of sociology or economics that it is one of the *behavioural* or *social sciences* can quite properly be taken to mean that it is an academic discipline whose subject matter is some aspect of human behaviour or some aspect of society.

In the nineteenth century, however, the primary meaning of the English word was narrowed down to its present use. The notion of a science came to be associated with the methods, concepts and credibility characteristic of the 'physical' and 'natural' sciences, whose progress was then so striking. Scholars opening up the systematic study of behavioural and social phenomena were induced to define their work as 'scientific' only when it followed those methods as far as possible. To say, then, that sociology is a social *science* could be interpreted as a claim that it shares some of the characteristics – especially the ultimate credibility – of physics, chemistry, biology or geology, and is capable of being applied technologically in the spirit of engineering or medicine.

The controversies surrounding this semantic confusion are obvious manifestations of *academic scientism* (§3.9). But underneath the rhetoric there lies the very important question of the influence of 'science' (in the narrower sense) within academia. In other words, we are led into an investigation of the extent to which other academic disciplines are – or ought to be – regarded as equivalent to, or comparable to, the established natural sciences. The object of such an investigation should not be to prove, say, that sociology is so 'like' physics that it could be just as 'true', but to discover what such disciplines have in common, and what they may learn from one another.

The precise extent of 'science' in the conventional sense is not quite clear. The natural sciences and their associated technologies (§9.6) do not constitute a compact set within academia. Disciplines such as geography, psychology and archaeology straddle the administrative boundaries between the Faculties of Science, Social Science and Humanities. Engineering, medicine and agriculture draw upon economics, social psychology and sociology as well as upon the physical, biological and earth sciences. Many practical techniques that originated in physics and chemistry, such as radiocarbon dating of manuscripts and works of art, have found invaluable application in the traditional humanities. Some branches of scholarship, by their very nature, have physical, biological and social aspects, whilst the technical resources of science can be exploited throughout the scholarly world, just as in everyday life.

But there has long been a tendency towards 'scientification', in many disciplines, that goes beyond mere eclecticism. For example, *quantitative* empirical data (§2.7) and quasi-mathematical theoretical *models* (§2.12) are often preferred over other forms of 'fact' and other schemes of explanation. The importance of *experiment* (§§2.8, 3.7) is also strongly emphasized, and the possibilities of successful *prediction* (§3.6) are considered. In other words, the methodologies and validation procedures that have proved so powerful in the physical sciences are being applied to behavioural and social phenomena, in the belief that they will produce a sounder body of knowledge than any alternative intellectual approach.

This belief is evidently founded upon a general philosophy of science that regards these as essential features of the 'scientific method'. But as the history of biology demonstrates, a vast amount of good science can be discovered with little reference to numerical measurements. The quantitative approach is not a necessary condition for epistemological progress in the scientific spirit, and tends to exclude qualitative information that is highly relevant to the understanding and explanation (§2.10) of complex phenomena. Again, direct experimentation may be feasible in the study of the behaviour of individuals and small groups, but is out of the question when the object of study is a whole social system. But natural scientists do not see this as a total ban of the logic of justification. For example, a conjecture in meteorology can be refuted by subsequent observations of uncontrived events, whilst the whole science of geology relies upon consistent theoretical inference (§3.5) from data obtained by systematic exploration of nature, with very little possibility of genuine prediction (§3.6). There is no absolute philosophical mandate for the proposition that science is essentially predictive, or that knowledge that cannot be tested experimentally must be excluded from the scientific archives.

Any opinion on the rôle of 'scientific method' in disciplines such as social psychology or economics thus depends on what is thought to be the nature of this method in its more conventional setting, and on the epistemological status of the results it obtains. Is scientific knowledge to be strictly limited to what can be discovered and validated by some formal procedure, such as the hypothetico-deductive method (§3.7) or does it include knowledge that could only have been obtained from other sources, such as introspective insight, or codified personal experience? Does science describe reality (§3.9) or is it only one of a number of possible ways of looking at the world?

Philosophy offers a variety of answers to these questions. Some philosophers, for example, maintain a strictly *positivist* epistemology (§3.3), which takes a very austere view as to what may be counted as scientific knowledge, but elevates this knowledge to a highly privileged status excluding almost all other sources. Supporters of scientific *realism* (§3.9) hold similarly that there is a uniquely credible scientific world view, which can be discovered by appropriate methods. General philosophies of this kind tend to make a sharp distinction between 'scientific' and 'non-scientific' procedures, and favour the introduction of the former in place of the latter in all branches of learning.

But these modes of *philosophical scientism* are now somewhat out of fashion, and science is usually assigned a more modest rôle in general philosophy. In particular, philosophers influenced by the *sociology of knowledge* (§8.1) would argue that all knowledge and belief must be thought of as a social product, whose character derives from the interpersonal situation where it is generated. Scientific knowledge should therefore be seen primarily as the collective output of a community with a peculiar internal social structure (§8.4) and subject to specific external forces. This critique does not necessarily lead to total relativist scepticism concerning the special status of scientific knowledge (§8.3), but it strongly suggests that the intellectual and personal relationships between the individual scholars in a particular discipline may be an indication of the extent to which it is 'like' an established science. Do they, for example, share enough by way of accepted facts and concepts to resolve some of their differences of opinion by rational argument, and thus extend the area of consensus (§8.5)? It was the lack of such efforts towards consensus between the warring schools of psychoanalysis that made the 'scientificity' of their enterprise seem so doubtful to outside observers.

In the face of such diversity of views about the true nature of science, it would be rash to express a firm opinion on its place within academia. Techniques and intellectual strategies developed in the natural sciences can certainly be employed profitably in the behavioural and social sciences. The 'scientific' attitudes of empiricism (§3.2), theoretical self-consistency (§3.8) and objectivity (§8.6) are always to be welcomed in any academic discipline. But every methodology of research has to justify itself by its results in action, and the modes of thought that have evolved in the study of physical and biological phenomena do not exhaust all our intellectual resources. The social sciences have had to develop their own characteristic strategies to deal with problems of enormous difficulty which have no analogues in the natural sciences. The philosophy of science has to be extended to deal with the *reflexivity* of social thought, which engulfs every social scientist in the action he or she is observing. Self-awareness is the primary human characteristic, and is not to be glossed over, or ignored, on the grounds that it is somehow reducible to more elementary behavioural components. Indeed the critique of positivism (§3.3) applies with even greater force in the interpretation of human behaviour than in physics or chemistry. Some of the epistemological problems that are obvious in attempts to construct 'scientific' theories of social phenomena may also be observed, in more insidious form, in the natural sciences. Instead of trying to make sociology look like physics, it may be wiser to accept that physics may not be so very different from sociology after all.

16.5 Science and values

From Plato, through Hobbes and Marx, one of the great dreams of European thought has been to construct a science of human behaviour that would solve all the problems of social life. In its grandest form, *political scientism* is akin to technocracy (§16.1). Politics itself is to be transformed into 'social engineering' to be blueprinted and operated by philosopher kings, scientific revolutionaries, or, more modestly, the diligent pupils of this or that professor of social science.

This conception of science as a complete cultural formula is untenable. The elusive art of government has undoubtedly gained immensely from the theoretical critiques and practical applications of the social sciences. But these critiques and applications are of limited scope, and are not founded securely on well-established theories comparable to the laws of physics and chemistry that have proved so effective in real engineering. As we have just seen, it is questionable in principle whether even the most mature and refined science of human behaviour could ever be precise and predictive to that degree. Moreover, any elaborate scheme of 'social engineering' would almost certainly fail in practice through lack of knowledge, or control, of numerous 'variables' and 'parameters' that could prove decisive in the unfolding of future events. Like every real technology, it would have to have a great deal of discretionary flexibility built into the design from the beginning.

Above all, any programme for the scientific management of human affairs is severely limited by uncertainties, contradictions and conflicts over the *values* at stake in the choice of objectives. These values derive from religious, ethical and aesthetic modes of thought and action that lie outside the domain of science as traditionally defined. In effect, political scientism violates the basic tenet of the academic ethos (§6.3) that science should be disconnected from political and religious causes.

Ever since the seventeenth century, scientists have proclaimed and celebrated the *neutrality* of scientific knowledge as a virtue associated with its objectivity and unimpeachable authority in its own sphere. The traditional philosophies of science have always followed the regulative principles of scientific work (§3.10) in insisting that the external world to be explored by science must surely have unique properties that are independent of the individual human mind. To undertake this exploration in a partisan religious or political spirit would be to blinker one's vision and risk failure in the search for truth. To draw organized science into the cauldron of politics (§ 14.6) – for example to enlist the Royal Society for or against the government of the day – would thus be a betrayal of the whole research enterprise.

Originally, neutrality of science had to be defined mainly in relation to *religion*. The delineation of this boundary was dramatized in the trial of Galileo by the Inquisition and in the public debates that followed the publication of Darwin's theory of evolution. Whatever the actual historical course of these particular events, they have proved potent myths in the establishment of the ideology of academic science. In each case, it later turned out that the region of knowledge that was to be taken over by science was not vital to religion. What could be clearly demonstrated, from publicly available empirical evidence, concerning the nature of the world in space, time and comprehensible pattern, could evidently be distinguished from the inspirational and ethical principles that people also need to order their personal lives. Religious and scientistic fundamentalists continue to dispute this boundary from either side, but most theologians and most scientists agree that there can be peaceful coexistence between science and religion, provided that they are not forced into direct confrontation.

Nowadays, the challenge to the neutrality of science comes mainly from politics. The collectivization of science (§11.1) has brought research under State control, so that decisions on science policy are inevitably influenced by political considerations (§14.5). At the technological end of the R & D spectrum, political and commercial forces flow into the gaps left by scientific ignorance and uncertainty, so that the scientific expert (§15.5) finds it almost impossible to maintain a neutral stance. In the basic natural sciences, it may look easy enough to draw a line between what *is* and what *ought* to be, between what *will* happen, and what is *desired* as the outcome of a certain action. But as the scientists who conceived and built the first nuclear weapons came to realize, there is no obvious frontier post between scientific *means* and political *ends* (§12.3). Perhaps the political neutrality of science was always a myth, as many Marxist metascientists have argued (§8.1): it is certainly unrealistic nowadays to suppose the pursuit of scientific knowledge can be disconnected from its political consequences and causes.

But the connections between the natural sciences and politics are explicitly technological. They are made through the instrumental mode, and have little to say on the religious, commercial, aesthetic or purely hedonistic goals that people seek by their means. A scientistic programme of 'social engineering' would have to employ the social and behavioural sciences to bring these values into the blueprint. But these sciences are certainly not as neutral and objective as this programme would demand. Every social psychologist, sociologist or economist, however disconnected his research from the corruptions of applicability, comes to realize that there is no such thing as a value-free proposition in relation to human affairs. The social sciences offer guidance on the likely outcome of social action as if through the eyes of an impartial observer; but this guidance, being expressed in the language and symbolism of a particular culture, already contains the values of that culture. In other words, the argument becomes circular. Science cannot be extended throughout the whole cultural domain to provide an independent objective source of both the means for action and the values attached to the results of such action.

16.6 The value of science

The fact that science cannot be the source for *all* human values does not mean that it cannot be considered a foundation for *some* values, or indeed, of value in itself. Many *rationalists* and *humanists*, having rejected the traditional religions, justify their ethical codes by reference to various fundamental scientific concepts, such as the apparent unity and coherence of the physical universe, or the inevitability of progress through biological evolution. Others are inspired by the notion of scientific technology as a means by which nature can be controlled and transformed. Others, again, revolt against this notion, and regard science as a major source of negative forces and values. The beliefs, hopes and fears generated by science are not, of course, part of science as such, and are not to be justified or dispelled by scientific analysis alone. These are themes that are often explored with imaginative insight through the genre of *science fiction*, where the social and cultural influence of science is often presented far more vividly and cogently than in the academic metascientific literature.

In all of this, one must not lose sight of the value attached to the pursuit of scientific knowledge as an end in itself. The unravelling of some great scientific mystery – for example, the decoding of the molecular mechanism of genetics – obviously gives enormous satisfaction to very large numbers of people, far beyond those who are directly involved. Quite apart from all utilitarian considerations, science is held in high esteem by the general public. The idea of science as a transcendental enterprise to explore the Universe, unveil the secrets of nature, and satisfy our boundless human curiosity (etc., etc: the rhetoric is also unbounded) is not a mere invention of the academic ideology (§6.4). For reasons that are none the less compelling because their ultimate sources are aesthetic and spiritual, people are willing to support basic science 'for its own sake', and take pride in scientific achievements whose significance they cannot properly understand. It may be our duty, in the field of science studies, to demystify scientific work, unmask the self-serving interests of scientists, devalue the products of social technology, and denounce the pretensions of science as a guide to social action. Nevertheless, when all the rhetoric has been debunked, there is a residuum of truth in the notion that science is a fascinating endeavour, capable of engaging men and women at their best, and enlarging and enriching the human spirit with its discoveries.

Further reading for chapter 16

The topics mentioned in this chapter are so large and diverse that it is inappropriate to recommend a limited set of sources. The following works happen to be relevant, and are amongst the many that a serious student would need to consult.

D. Layton, 'Education in Industrialized Societies', in *A History of Technology*, ed. T. I. Williams, Vol. VI, Part I, pp. 138–71. Oxford: Clarendon Press, 1978

J. R. Ravetz, *Scientific Knowledge and its Social Problems*. Oxford: Clarendon Press, 1971 (pp. 364–402 on 'Folk Science')

R. Westrum, 'Science and Social Intelligence about Anomalies: The Case of Meteorites', in *Sociology of Science Knowledge*, ed. H. M. Collins, pp. 185–217. Bath: Bath University Press, 1982

L. R. Graham, *Between Science and Values*. New York: Columbia University Press, 1981

T. Roszak, *Where the Wasteland Ends*. London: Faber & Faber, 1973 (an attack on scientism from a humanistic point of view)

J. Passmore, *Science and its Critics*. London: Duckworth, 1978 (a defence of science as a component of culture)

Index

abstract journals, 59
academic science, 5–11, 105–6, 119, 139
  communication system, 65, 68
  ethos, 85, 87
  external relations, 126–7, 156, 164
  formal institutions, 93
  historical origins, 122–8
  instrumental rôle, 132–3, 135, 141
  social structure 70, 72, 75, 77, 81, 99–100, 146, 174, 178
academies, national, 7, 66, 71, 73, 75–6, 124, 159, 170, 180
accelerator, particle, 49, 78, 135, 138, 161, 165
accountability, 145, 160, 164–5
accountancy, financial, 154–7
accreditation of knowledge, 65–7, 71, 73–4, 86, 93, 186
administration, 72, 78, 85, 123, 128, 141, 145, 160, 165, 170, 176–7
Advancement of Science, Associations for, 126
adviser, 179, 181
aggregation of facilities, 136–7
agriculture, 113–14, 129, 133, 150–1, 164, 168, 186
aircraft industry, 156
amateur scientist, 16, 123–4, 134
analogy, 25, 27, 30, 54, 97
anatomy, 15, 18–19, 26, 97
Anderson, P. W., 60
Andrews, K., 172
anomaly, 26, 63, 95, 98
anthropology, social, 72, 185
antibiotics, 150
apparatus, 22, 38, 74, 134–5, 149, 175
applied science, 129–30, 132–3, 138, 141, 144, 165
Archimedes, 20
architecture, 114, 119
archives, scientific, 1, 2–3, 10, 58–9, 62, 67–8, 81, 91, 102, 108, 110, 114

Arrow, K. J., 157
assistant, research, 134, 136–8, 175
astronomy, 19–20, 22–3, 46, 93, 95, 126–7, 133–4, 137, 186
atom, 18, 25, 27, 30, 39, 40, 51
Atomic Energy Research Establishment, 142
attitude, scientific, 18, 88, 106, 175, 184, 190
author, 59, 62–5, 75, 77, 81, 138, 178
authority, managerial, 144–7, 160, 170
  scientific, 8, 9, 70–81, 84, 87, 94, 100, 125, 176, 181
automobile industry, 114–15, 130, 156
autonomy, communal, 118, 128, 139, 159, 171
  individual, 88–9, 125–6, 147, 164–5, 178
axiomatization, 50, 53
Azande, 104

Babylonian science, 23
Bacon, 113
Barber, B., 33
Barnes, B., 57, 80, 111, 120, 181
basic research, 3, 14, 113, 141–3, 145, 152–4, 193
  funding, 156, 163–5, 168, 175
Bayes' Theorem, 45
behavioural sciences, 6, 36, 109, 187–90, 192
behaviourism, 39
belief, 5, 43, 68, 105, 109, 186–7
Bell Telephone Laboratories, 128
Ben-David, J., 131
Bernal, J. D., 11, 120
Beveridge, W. I. B., 33
bibliography, 59
big science, 138, 160
biochemistry, 26, 43, 94, 114, 132, 186
biography, 29, 70, 92
biology, 6, 15–17, 50, 103, 173, 188
biomedical research, 133, 167
biotechnology, 94, 113, 142, 155, 167
Blisset, M., 148

195

Index

Bloor, D., 111
Böhme, G., 101, 120
Bohr, 27
books, 58, 67, 92, 135
botany, 14, 16, 20, 126
Boyle's Law, 23
Bradbury, F. R., 157
Brahe, Tycho, 134, 137
Britain, 124, 125, 131, 133, 153, 159, 170–1
Broad, W., 90
Brooks, H., 172
bureaucracy, 78, 81, 89, 125, 128–9, 145, 165, 170
Bush, V., 133, 139
Butterfield, H., 11

caloric, 34, 96
Campbell, N., 57
cancer, 26, 32, 142–3, 161, 168, 186
capitalism, 103–4, 116, 127, 149, 155–7, 160, 167, 175
Cardwell, D. S. L., 131
Carnot, S., 117
Casimir, H., 91
cause, 25–6, 42, 116
Cavendish Laboratory, 135
Caws, P. 56
Ceci, S. J., 69
CERN, 177
change, scientific, 53, 68, 78, 91–102
chemical industry, 114, 127, 133
chemistry, 6, 13, 17, 20, 22–4, 27, 30, 38, 45, 50, 51, 59, 89, 93, 96–7, 144
China, 122, 134
choice, criteria for, 161–4, 172
chromatograph, gas, 137
citation, 59–61, 63, 65–7, 70, 74–5, 77, 81, 84, 94, 166
cladistics, 16
Clark, J. 158
classification, 15–18, 20–1, 23, 37, 59, 116
cognition, 33, 109
Cole, J. R., 80
Cole, J. S., 80
collaboration, 76, 136–8
collectivization of science, x, 5, 130, 132–9, 160, 165–6, 170, 174, 176, 191
Collins, H. M., 33, 57, 111
commerce, 5, 71, 86, 126, 128, 137, 140, 142, 179
common sense, 35, 52
communalism, 84, 86, 89, 107, 146, 175
communication, scientific, 10, 56, 81, 84, 107–8, 144, 176
  informal, 67–8, 73, 75, 82, 84, 92
  system, 58–68
communism, 169–70

community, scientific, 3, 6, 8, 10
  growth of, 123
  international, 71, 124, 177
  rules, 81–2, 174
  social rôle, 126, 128, 139, 169–71, 180
  structure, 60, 70, 72, 75–6, 78–9, 107, 129, 176
competition, 63, 72–4, 76, 81, 85, 88, 108–9, 125, 137, 144, 175
computer, 27, 59, 115, 136
concept, 24, 28, 30, 46, 49–50, 53, 129, 143
conferences, 67, 71–2, 75, 88, 94
confirmation, 38, 40, 44, 46
conjecture, 11, 30, 35, 47–8, 51–3, 63, 93, 96, 98, 103, 188
consensibility, 109
consensus, scientific, 10–11, 56, 62, 68, 78, 89, 108–9, 189
constructivism, 54
consultant, 85, 126, 179
continental drift, 46, 51, 66, 82, 97–8, 187
contingency, 37
control, social, 118
  of science, 161, 166–9, 172, 191
controversy, 6, 8, 16, 66, 85, 97, 109, 179
conventionalism, 52–5
correlation, statistical, 26, 42
corrigibility, 34, 50–2
corroboration, 45–7, 51, 95–6, 143
cosmology, 5, 44, 95, 115, 122
craft, 115–16, 119, 129
Crane, D., 80, 100
creativity, 29, 85, 92
Crick, F., 163
criticism, 62, 65–6, 73, 76, 81, 85, 91, 95, 178
Crossfield, K., 157
CUDOS, 86–7
culture, science in, 183–93
curiosity, 2, 19, 48, 87, 174, 193
Czechoslovakia, 125

Daele, W. van den, 120
Darwin, 25, 58, 97, 103, 191
data compilation, 59, 135, 188
de Mey, M., 33
deduction, 40, 46
Defense, Department of, 133
demarcation principle, 14, 47, 89
democracy, 79, 81, 127, 145, 169–70, 181
description, 14–15, 40
determinism, 55
developing countries, science in, 67, 115, 152, 160, 167
development, technological, 85, 121–3, 129, 132, 138, 142, 146, 152–4, 157, 162
Dingle, H., 82

discipline,
  academic, 6, 58, 74, 94, 98–9, 144, 188
  metascientific, ix, 1, 2, 6, 9–10
disconfirmation, 40, 45, 47, 54, 61, 96
discovery, 2–3, 7, 10, 13, 22, 32, 54, 66, 72, 85, 92–3
  claim, 60, 72, 85
  context of, 13, 18, 48
  erroneous, 34
  simultaneous, 73, 99–100
disinterestedness, 62, 85–7, 146, 175, 180
division of labour, 32, 74, 138
DNA, 27, 73, 93, 97, 163
Dupree, A. H., 131
dynamics of science, 48, 79, 86, 99–100

economics, 97, 138, 140, 187, 189, 192
  of R & D, 1, 3, 5, 70, 134, 140–1, 149–57, 160, 166
Edge, D., 57, 80, 120, 181
Edison, T., 113
editor, 63–5, 71, 75, 78, 81, 176
education, ix, 10, 94, 106, 125–6, 128
  science, 2, 6, 22, 67, 86, 97, 144, 184–5, 193
effect, 25–6, 116
egoism, 175
Einstein, A., 29, 34, 39, 46, 49, 70, 72–3, 87, 95, 98, 184
electrical industry, 113, 117, 133, 150
electromagnetism, 29, 49, 96, 150
electronics, 127, 144, 154, 156
Elias, N., 80, 148
emergence, 55
empiricism, 22–4, 28, 31, 35–8, 40, 47, 50, 52, 110, 143, 190
employment, 71, 125, 128–9, 146, 159, 177–8
encyclopaedias, 13, 66
energy, 52, 112, 114, 117, 142, 160, 168–9
engineering, 6, 27, 32, 94, 113–14, 116, 119, 122, 142–3
  profession, 126–7, 129, 152, 178
  social, 190
entrepreneur, scientist as, 174–6, 178, 180
environment, 146, 164, 168
epistemology, 9, 10, 34–5, 40, 54, 68, 103, 109–10, 126, 129, 137, 189
eponymy, 70
error, 20, 34, 53, 63, 65, 85, 93, 103–4
establishment, 79, 90, 97, 165, 170–1
ethics, 143, 180, 191–2
ethnographic approach, 7, 68, 83
ethos, academic, 86–9, 92, 107, 125–7, 129, 138, 144, 147, 163, 173, 175–6, 180
Europe, 122, 175
Evans, W. G. 158
event, 18, 39

everyday world, 20, 35–8, 49, 52, 54, 118, 126
evidence, 20, 26, 36, 40, 43, 180
evolution,
  biological, 8, 25, 51, 55, 58, 97, 103, 191–2
  epistemological, 51, 56, 68, 93, 99
  technological, 155
exchange model of science, 72, 139, 175
experiment, 1, 22–3, 31–2, 37, 44, 46, 48, 51, 116, 137–8, 143, 188–9
expert, 71, 85, 94, 141, 178–81, 191
expertise, 31, 89
explanandum, 25, 28
explanans, 24
explanation, 24–8, 39–40, 44, 54, 56, 62, 91, 115, 188
exploratory research, 18, 48, 54, 63, 126, 166, 188
extrasensory perception, 105, 187

fact, scientific, 14, 16–17, 19–20, 24, 27, 31, 35–9, 43, 48, 50, 54, 98, 110
falsification, 45, 47, 88
Faraday, M., 29, 92, 102, 113, 149
Fermi, E., 113
Feyerabend, P., 57, 101
finalization, 118, 144
Fisch, R., 181
Fleck, J., 80
Fleck, L., 99–102, 110
float glass, 152
folk science, 173, 186, 193
formalism, 29–30
foundation, charitable, 136
Fox, R. C., 33
France, 124, 159, 171
Franklin, R., 82
fraud, 82–3, 90
freedom in science, 118, 125, 178
Freeman, C., 157–8
Friday, 107
functionalism, 11, 78, 83
fundamentals, 24–5, 54, 56, 142, 191–2
funding, research, 140
funding agencies, 76, 165, 175

Galileo, 19, 121, 170, 191
Garfield, E., 69
gene, 25, 28, 35, 53
General Electric Company, 127, 133
generalization, 15–16, 23–5, 40, 42–4, 47, 54, 62, 108
genetics, 27, 86, 93, 143
genus, 16
geology, 14, 16–19, 22, 44, 46, 53, 97, 127, 138, 188
geometry, 53
Germany, 124–5, 131, 153, 159

Index

Gibbons, M., 158
Gilbert, W., 13
GNP, 151–2
government, 71–2, 106, 126, 133, 139–40, 157, 169–72, 190
governmental science, 127, 131, 145, 159, 164, 168
Graham, L. R., 193
grant, research, 71, 76, 86, 146–7, 165
gravitation, 25, 46, 51, 95
Greenberg, D. S., 139
Griliches, Z., 150, 157
growth,
  economic, 151–2
  of science, 123, 136
Gummett, P., 172

Haberer, J., 181
Hagstrom, W. O., 80
Hall, A. R., 11
Harnad, S., 69
Harré, R., 56
Heath, J. B., 157
Heisenberg, W., 159
Helmholtz, H., 183
hermeneutics, 7
Hesse, M., 57
Hessen, B., 116–17
Hindsight, 153
Hirsch, W., 33
Hirschman, D., xi
history,
  of science, 1, 3, 4, 6, 8, 11–12, 14, 29, 30, 34, 67, 91–100, 102–3, 123–8, 134, 175
  of technology, 113, 116, 122, 150
Hobbes, T., 190
holism, 55
Holland, 124
Holton, G., 30, 33
honesty, 83, 85, 88, 174
Hooke, R., 124
humanism, 192
humanities, 10, 13, 58, 89, 108, 188
Hume, D., 40–1, 47, 105
Hungary, 125
hybrid corn, 150–1
hydrology, 138
hypothesis, 29–31, 35, 38–9, 42–4, 46, 51, 98, 116, 130
hypothetico-deductive method, 46–8, 51, 86, 88, 91, 93, 130, 189

idealism, 55
ideology, academic, 87–90, 100, 108, 126, 129, 147, 165, 191, 193
  political, 16, 89, 103, 112, 127, 166, 175
imagination, 29, 48, 85, 99, 122

immunology, 8, 35
incommensurability, 96, 104
incorporation of science, 170
India, 134
indicators, science, 166
individualism, 73, 88, 92, 125, 138, 143, 163, 166, 171, 173–5
induction, 40–5, 47, 103, 105
inductivism, 41
industrial science, 122, 127–30, 132, 138, 141, 145, 177
industrialization of science, 132, 139
industry, 3, 63, 71, 79, 119, 126–9, 136, 139, 140, 142, 152, 160, 170
inference, 41–5, 52, 188
information, 46, 58, 65, 91
innovation, technological, 117–18, 143, 151, 152–3, 155, 168
institutions, scientific, 6, 58, 71, 81, 93–4, 100, 124, 135, 144, 180
instrument, scientific, 19–21, 36, 38, 46, 63, 91, 108, 134, 136–7, 144
instrumental conception of science, 1, 2, 5, 112, 115, 119, 128–9, 139, 140–2, 167, 169, 183–5, 192
intelligence, 2, 173–4
interdisciplinary research, 144
International Geophysical Year, 177
internationalism, 71, 124, 177
interpretation, 29, 143
intersubjectivity, 36, 109–10
intuition, 2, 29, 45, 109
invention, 117, 123, 128, 134, 141–2, 152, 155
investigation, 18–19, 31, 62
investment, 151, 153–6, 161
invisible colleges, 75–6, 78, 81, 94, 99–100, 163, 166, 177

Jagtenberg, T., 172
jet engine, 153
Jevons, F. R., 158
Jewkes, D., 157
Johnston, R., 172
Joule, J. P., 117
journals, scientific, 6, 13, 58–9, 63, 75, 94
justification, context of, 14, 32, 35, 37, 48

Kelvin, *see* Thomson, W.
Kepler, J., 25, 137
Knorr-Cetina, K. D., 69
knowledge, 13, 28, 68
  accumulation of, 1, 4, 18, 51, 60, 74, 92, 103, 121
  established, 48–51, 54, 65, 94, 97, 108, 143, 163, 179, 186
  growth of, 31, 54, 68, 79, 91
  manufacture of, 106

Knowledge (cont.)
  organized, 2, 92
  practical, 114–15, 119, 185
  public, 8–11, 56, 58, 68, 84, 107–9, 153–4, 184
  tacit, 39, 42, 47, 54, 68, 78, 107, 115, 129, 179
Kondratiev, N., 151
Körner, S., xi
Krohn, W., 120, 172
Kuhn, T., 94–100, 102, 104, 110

laboratories, 135, 149
laetrile, 186
Lakatos, I., 96, 101
Lakoff, S. A., 181
Lamberton, D. M., 157
Landau, L. D., 77
Langmuir, I., 133
Langrish, J. 158
language, 15, 30, 62, 108
  observational, 38
  theoretical, 38
laser, 113
Latour, B., 12
Laudan, L., 101, 111
Lavoisier, 58
law, 10, 13, 24, 36, 83, 107, 112, 124, 169, 179–80
  scientific, 23–6, 28, 40, 42, 44, 46, 108
Layton, D., 193
Layton, E. T., 120, 158, 172
Leeuwenhoek, 124
Leonardo da Vinci, 107
libraries, 59, 63, 110, 135
life-world, 49, 52, 110, 115, 185
linguistics, 110
literature, scientific, 58, 68, 74, 84
  primary, 58–60, 62–3
  secondary, 58–9, 66
logic, 2, 25, 28, 40, 45–6, 50, 55, 62, 108–9
Lotka's law, 77
Lysenko, T. D., 87, 186

Mach, E., 39
MacLeod, R., 12
macroeconomics, 150–3
magic, 186
Mahoney, M., 90
Malthus, 103
management, 72–3, 78, 126, 128, 139
  of R & D, 145–7, 155, 162, 166, 170, 177–8
map, cognitive, 49, 53, 56, 74, 94, 185
Marcson, S., 148
Martins, H., 80, 148
Marxism, 84, 112, 116, 118, 166–7, 190–2
mass spectrometer, 20, 137

materialism, 52–3
  dialectic, 116
materials science, 114, 138
mathematics, 18, 21, 24–5, 27–8, 30, 43, 53, 95, 108, 144
  pure, 5, 29, 50, 133, 136, 145
Matthew effect, 77, 84, 166
Maxwell, J. C., 29, 49, 81
Mayr, O., 120
Meadows, A. J., 69, 79, 100
measurement, 20–1, 24, 31–2, 46, 188
mechanics, 24–5, 51, 96, 98
mechanism, 27
Medawar, P., 8
media, mass, 71, 184
medicine, 6, 25, 83, 104, 113–14, 121–2, 124, 126, 133, 138, 143, 150, 186
meetings, 67, 84, 86, 176, 178
Menard, H. W., 80, 100
Mendel, 28
Mendeleev, 23–4, 43, 45
merit,
  personal, 77, 84
  scientific, 163
  social, 162–3
  technical, 162–3
Merton, R. K., 80, 83, 86, 89–90, 131, 147, 176
metallurgy, 114
metaphor, 25, 30, 49, 54, 96, 104, 175
metaphysics, 38–9, 47, 55, 108–9
metascience, x, 1–3, 5, 7, 9–10, 96, 102, 105, 110, 130, 149, 161, 166, 173, 191
meteorites, 36, 187, 193
meteorology, 27, 117, 188
method, scientific, 1–3, 7, 10, 13, 22, 32, 35, 43, 47, 60, 62, 68, 93, 103, 105, 107–8, 115, 141, 143–4, 188–9
methodology, 15, 21, 22, 95, 98, 129, 190
Metzger, W. P., 181
microbiology 19, 97, 137
microeconomics, 154–7
microscope, 19, 38
  electron, 137
military R & D, 5, 86, 122, 128, 146, 150, 152–3, 160, 164, 167, 169, 172, 177–8, 180–1
military technology, 117, 119, 134–5, 156–7, 169
military–industrial complex, 79, 139, 171
mining, 114
mission-oriented research, 142, 146, 150, 153–5, 162
model,
  of science, 2–3, 10, 72, 95–7, 106–10, 127
  scientific, 26–30, 46, 50, 53–4, 188
molecular biology, 28, 35, 48, 50, 73, 93–4, 97, 119, 143–4, 192

monograph, 66
Mulkay, M., 69, 90, 110
multinational corporation, 155, 160
Musgrave, A., 101

Naess, A., 57
National Bureau of Standards, 133
National Institutes of Health, 133
National Physical Laboratory, 127, 133
National Science Foundation, 133
nationalism, 84, 159, 170, 177
natural history, 14, 16, 103
natural sciences, 6, 11, 14, 102, 106, 130, 187, 189–90
Nature, 8, 63
needs, human, 140, 157, 167
Nelson, R., 157
Nelson, W. R. 139
network,
  co-citation, 60–1, 75, 94
  communication, 67
  conceptual, 49, 59, 163, 168
von Neumann, J., 140
neutrality, 1, 180, 191–2
neutrino, 52
Newton, I., 24–5, 46, 51, 95–8, 104, 117, 170
Nobel prize, 60, 66, 71–4, 76
normal science, 95–9
norms, of science, 3, 6, 9, 81–90, 100, 105, 107, 122, 125, 129, 138, 145, 166, 174
Norris, K., 157
NRDC, 155
nuclear engineering, 112–13, 117, 134, 138, 146, 168, 179
nuclear weapons, 26, 73, 117, 132, 135, 173, 191
nutrition, 114
nylon, 154

objectivity, 2, 6, 7, 36, 62, 68, 85, 88, 92, 104, 108–10, 146, 179, 190
observables, 38, 50
observation, 1, 10, 15, 18–22, 31–2, 37, 61, 134, 137, 143
observer, 20, 36, 109
oligarchy, 78–9, 125, 139, 145
ontology, 52, 54
operationalism, 39
Oppenheimer, J. R., 72, 173, 183
opportunism, 34, 83
order, taxonomic, 16
ordering principle, 16–18, 27–8, 55–6, 88
originality, 8, 22, 60, 63, 65, 73–4, 85–7, 91, 98, 100, 108, 137, 147, 166
ornithology, 15, 17
Oteri, J. S., 181

palaeontology, 15, 44, 48

paper, scientific, 8, 58, 61–4, 70, 74, 81, 106, 138
paradigm, 95–100, 104, 110, 118, 136, 143–4, 163
parascience, 105–6, 187
Passmore, J., 193
Pasteur, L., 92
  Institut, 135
patents, 84, 152, 154
pathology, 26
patronage, state, 124, 134, 159, 164–6
pattern, 16–18, 21, 23, 26, 29, 55
peace, 171, 177
peer review, 11, 64–5, 75–6, 86, 136, 147, 165–6
penicillin, 132
perception, 19, 36, 39, 109
periodical, scientific, 58
persistence, 174
personality, 173–6, 178, 184
Peters, D. P., 69
pharmaceutical industry, 156
pharmacology, 119
phenomena, 22, 37–40, 44, 130
phenomenalism, 39
phenomenology, 23–4, 28, 40, 44, 50, 110
philosophies of science, 52–5, 86, 91–3, 96, 115, 179, 189–91
philosophy of science, xi, 1–5, 7–9, 13–57, 89, 102, 107, 112, 126, 130, 143, 186
phlogiston, 96
photography, 36, 38
physiology, 26, 43, 97, 114
physics, 6, 17–19, 21, 27, 39, 49, 50, 58, 73–4, 89, 93, 117, 129, 132, 137–8, 144, 154, 179, 188, 190
  classical, 95–6, 153
  elementary particle, 18, 22, 35, 49, 138, 143
  theoretical, 48, 50, 51, 53, 75, 103, 142, 173
Pinales, M. S., 181
Pinch, T. J., 111
plagiarism, 85
plasma physics, 142
plate tectonics, 14
Platonism, 34, 190
platypus, 36
Pledge, H. T., 11
pluralism, 89, 118, 127, 166, 180
pointer readings, 19, 38
Poland, 125
Polanyi, M., 33, 57, 89
politics,
  science and, 1, 2, 5, 13, 96–7, 118, 141, 160–1, 168–9, 190–2
  scientists in, 72, 78, 138, 141, 179, 183–4
Poole, J. B., 172

Popper, K. R., 45, 47, 56–7, 96, 98, 108–9, 111
popularization of science, 126, 184
positivism, 34, 39, 43, 54, 189
potlatch, 72
Potter, J., 69
Powers, J., 57
practice, 126, 143, 167, 179
pragmatism, 44, 48–9, 110, 117, 122, 144
prediction, 43–7, 51, 115, 118, 143, 188–9
pre-prints, 67, 75, 94
Price, D. de Solla, 12, 80, 100–1, 120, 157–8, 172, 181
Price, D. K., 172
priority of discovery, 59–60, 85
prizes, 6, 66, 71, 76, 84–5, 88, 129
probability, 41–3, 179
problem solving, 1, 2, 31–2, 48, 66, 74–5, 91–3, 95, 97–9, 130, 137, 140, 147, 168, 179, 190
productivity, 63, 77, 145
profession, scientific, 2, 63, 70, 79, 81, 83, 93–4, 125–6, 129, 174, 177–8
profits, 140, 149–50, 154, 156–7
programme, research, 6, 96, 99, 101, 105–6, 163
progress, 56, 60, 92–3, 100, 103, 118, 123, 137, 140, 166–7, 192
pseudo-science, 186–7
psychoanalysis, 189
psychology, 39, 83, 110, 189, 192
  of research, 1–5, 7–9, 19, 70, 85, 92, 109, 130, 173–4
publication, 8, 62–4, 76, 84–5, 107, 166
publishing, scientific, 63, 93–4
Pugwash meetings, 177
pulsars, 37
pure science, 5, 129–30, 132–4, 138, 141, 159, 165, 180
puzzle, 32, 95, 97–8, 100
Pythagorean, 34

quality, 62–3, 72–3
  of life, 140
quantum mechanics, 21, 24, 49, 51, 53, 113
question, scientific, 18, 31, 61, 98

R & D, 121–3, 132, 139, 143–5, 151, 167–8, 170–1
  organizations, 5, 122, 133, 140–8, 152, 160, 177–9
racialism, 84, 89
radar, 113, 135
radioastronomy, 19, 37, 135, 138
rationalism, 55, 192
rationality, 28, 31, 41, 43–4, 46, 62, 88, 93, 104, 106–7, 122, 143

Ravetz, J. R., 33, 132, 139, 193
realism, 34, 49, 52–5, 88, 105, 110, 143, 189
recognition, 61, 70–4, 81, 85, 94, 100, 108–9, 125, 129, 138–9, 144, 170, 174
reductionism, 38, 43, 55–6
referee, 11, 64–5, 71, 76, 81, 84, 165, 176
reference, bibliographic, 59
reflexivity, 190
refutation, 47, 93, 96, 98, 103, 188
regulative principles of science, 55–6, 78, 88–9, 92, 104, 109, 115, 130, 143, 186, 191
Reid, R. W., 182
relativism, 54, 103–5, 109–10, 187, 189
relativity, theory of, 21, 29, 39, 46, 49, 73, 95–6, 98
relevance, 29, 122, 141–3, 146, 162, 185
religion, 10, 13, 84, 89, 91, 104, 107, 112, 124, 169, 175, 185–6, 191–2
replication of experiments, 22, 37–8, 88
reproducibility, 22, 37, 109
republic of science, 81, 139, 176–7, 180–1
research, 2, 13–33, 56, 121–2, 132
  commissioned, 85, 147, 168
  industrial, 5, 85, 127–9, 135
  (see also basic research)
research council, 119, 133, 140, 165, 168
responsibility, social, 126, 176, 180–1, 186
retrieval of information, 58–9
review articles, 59, 66
reviewer, 64
revolution, scientific, 51, 53, 91, 94–9, 102, 143, 175,
  technological, 118
Richter, M. N., 172
rights, human, 89, 177
rôle, of scientist, 5, 6, 88, 107, 125, 141, 174–80, 183–4
Rose, H., 120, 139
Rose, S., 120, 139
Rosenberg, N., 158
Roszak, T., 193
Rotblat, J., 182
Rothschild Report, 168
Royal Aircraft Establishment, 128
Royal Society, 76, 79, 81, 121, 124, 159, 170, 191
rules of scientific behaviour, 9, 81–3, 107, 174
Russia, 73, 87, 124, 134, 146, 159, 171, 175, 186
Rutherford, E., 112–13, 132
Ryle, M., 19

S & T, 119, 121–3, 169
Sakharov, A., 73
Salomon, J. J., 148
Sapolsky, H. M., 172

Sawers, D., 158
scepticism, 18, 35, 39, 54, 85–6, 88, 91, 97, 105, 108, 122, 147, 166, 178, 186
Schmookler, J., 157
Schumpeter, J., 151
Science, 8, 63
science fiction, 192
science policy, 133, 141, 143, 160–9
scientism, 53, 105, 183–4, 187–91, 193
secrecy, 84, 86–7, 128, 146, 154, 178
sense-data, 39
serendipity, 18, 33, 48, 118, 143, 166, 174
Shinn, T., 148
Siemens, W., 113
simplicity, 25, 34, 53
Singer, C., 11
Sklair, L., 12, 90, 148
social sciences, 6, 7–8, 14, 36, 58, 82, 89, 102, 106, 109, 168, 187–90, 192
socialism, 160, 167
societies, learned, 6, 63–5, 71, 75, 78, 81–2, 93–4, 129, 170, 176–7, 180
sociology, 187–8, 190, 192
  external, 3–6, 10, 112, 118, 122, 126, 130, 139, 149, 160
  internal, 3–6, 10, 64, 67, 76, 82–5, 87, 89, 99, 107, 122, 128, 130, 138, 145–7
  of knowledge, 5, 54, 68, 78, 102–10, 144, 166, 180, 189
  of science, 1, 3, 7–9, 89, 100, 105, 147, 160
Soete, L., 158
solipsism, 35, 55
sophistication, instrumental, 136–7
space research, 138
specialization, 74, 141, 163, 175, 178, 184
specialty, scientific, 30, 49, 61, 63–4, 74–5, 78, 84, 94, 98–100, 138, 168
species, 16, 37
spectrometry, 137
speculation, 32, 95, 97
Spiegel-Rösing, I., 12, 101, 157–8, 172, 181
spontaneous generation, 34
Starnberg hypothesis, 144
State, 124–6, 176
  support of science, 133, 136, 140, 159–60
statistical analysis, 41–3
steam engine, 116–17
Steele, E., 8
Stillerman, R., 158
Storer, N. W., 90
strategic research, 142, 145–6, 150, 153–4, 156, 163, 168
stratification, 76–9, 84, 99, 134
subjectivism, 55
subjectivity, 6, 20, 36, 109
Succession, Law of, 41, 44
superstition, 105

Switzerland, 124
symbol, 49, 53

talent, 77
targetted research, 142, 153
taxonomy, 15–18, 28, 37, 59
team research, 138, 143, 146–7
technique, scientific, 114–15
technocracy, 184, 190
technology, 1, 2–5, 10, 32, 51, 89, 93, 102, 112–20, 142, 161, 185, 187–8
  assessment, 169
  science-based, 113–15, 117, 119, 127, 144, 156–7, 161, 184
  transfer, 152
technology-based science, 114, 117, 119
telegraphy, 118
telescope, 19–20, 38, 134
television, 118
tenure, academic, 71, 76, 81, 125, 144, 175
textbooks, 13, 66, 73
themata, 30, 102, 117
theorizing, 1, 28, 31–2, 40, 48, 95, 143
theory, 24, 28–32, 38, 40, 45, 48, 54, 98, 129
thermodynamics, 24, 114, 143, 153
Thomson, W., 117, 126
thought collective, 99–100
tomography, 19
Toulmin, S., 101
TRACES, 153
tradition, 6, 99
transistor, 154, 156
Trend Report, 133
trial and error, 32, 47, 116, 119, 122, 130
Tricker, R. A. R., 57
truth, 34, 43, 51, 55, 89, 105, 110, 191

understanding of science, 184–5, 193
unidentified flying objects, 106
United Nations, 177
United States, 124, 131, 133, 139, 150–1, 160, 170–1, 175
universalism, 84, 87, 89, 146, 186
universities, 63, 72, 93, 123–5, 128, 133, 135–6, 140, 142, 171, 180
University Grants Committee, 133
utilitarianism, 133–4, 139, 142, 185, 192

Vaizey, J., 157
validation, 2, 10, 24, 32, 34–57, 65, 73, 85, 103–4, 107, 116
values, human, 143, 190–3
Velikovsky, I., 186
vocation, scientific, 2, 5, 10, 174

Wade, N., 90
Wallace, A. R., 103

war, 140
Watson, J., 82, 163
waves, 27, 29
  long, 151
Weber, M., 144
Wegener, A., 97–8
Weinberg, A., 148, 162, 172
Weinberg, M. G., 181
Weingart, P., 80, 120, 172
Werskey, G., 120
Westrum, R., 193
whistle-blowing, 147, 181
Whitley, R. D., 80, 148
Williams, L. P., 149
Williams, T. I., 157, 193
Woolgar, S., 12, 33

workers, scientists as, 177–8, 181
'world 3', 108–11
world picture, scientific, 185, 189, 191
World War, 177
  Second, 128, 132, 135, 153, 170

xerography, 152

Yearley, S., 69
Yoxen, E., 80

Zen, 109
Ziman, J. M., 12, 90, 111, 148
zoology, 14, 16
Zuckerman, H., 79